D0204591

Cities, Politics and Power

Traditionally, the study of 'power in the city' was confined to the institutions of urban government and the actors involved in contesting and making political decisions in and for metropolitan societies. Increasingly, however, attention has turned to the function of the city not only as a centre of urban governance but as a major economic, social, cultural and strategic force in its own right.

Cities, Politics and Power combines this traditional concern with how the cities in which we live are organised and run with a broader focus on cities and urban regions as multiple sites and agents of power. This book is divided into five parts, with a short introduction outlining the argument and organisation of the text. Part II charts the development of the urban polity and considers the ways in which coercion and force continue to be used to segregate, oppress and annihilate urban populations. Part III critically examines the key collective actors and processes that compete for and organise political power within cities, and how urban governance operates and interacts with lesser and greater scales of government and networks of power. Part IV then explores the ways in which 'the political' is constituted by urban inhabitants, and how social identity, information and communication networks, and the natural and built environment all comprise intersecting fields of urban power. The conclusion calls for a broader theoretical and thematic approach to the study of urban politics.

This book makes extensive use of comparative and historical case studies, providing broad coverage of politics and urban movements in both the Global North and the Global South, with a particular focus on the UK, USA, Canada, Latin America and China. It is written in an accessible and lucid style and provides suggestions for further reading at the end of each chapter.

Simon Parker is Senior Lecturer in Politics at the University of York where he teaches urban theory and comparative politics. He is the author of *Urban Theory and the Urban Experience: Encountering the City* (Routledge, 2004).

Routledge critical introductions to urbanism and the city

Edited by Malcolm Miles, University of Plymouth, UK
and John Rennie Short, University of Maryland, USA

The series is designed to allow undergraduate readers to make sense of, and find a critical way into, urbanism. It will:

- Cover a broad range of themes
- Introduce key ideas and sources
- Allow the author to articulate her/his own position
- Introduce complex arguments clearly and accessibly
- Bridge disciplines, and theory and practice
- Be affordable and well designed.

The series covers social, political, economic, cultural and spatial concerns. It will appeal to students in architecture, cultural studies, geography, popular culture, sociology, urban studies and urban planning. It will be trans-disciplinary. Firmly situated in the present, it also introduces material from the cities of modernity and post-modernity.

Published:

Cities and Consumption – Mark Jayne
Cities and Cultures – Malcolm Miles
Cities and Nature – Lisa Benton-Short and John Rennie Short
Cities and Economies – Yeong-Hyun Kim and John Rennie Short
Cities and Cinema – Barbara Mennel
Cities and Gender – Helen Jarvis with Paula Kantor and Jonathan Cloke
Cities and Design – Paul L. Knox
Cities, Politics and Power – Simon Parker

Forthcoming:

Children, Youth and the City – Kathrin Hörshelmann and Lorraine van Blerk
Cities and Sexualities – Phil Hubbard

Cities, Politics and Power

Simon Parker

Routledge
Taylor & Francis Group
LONDON AND NEW YORK

First published 2011
by Routledge
2 Park Square, Milton Park, Abingdon, Oxon, OX14 4RN

Simultaneously published in the USA and Canada
by Routledge
270 Madison Avenue, New York, NY 10016

Routledge is an imprint of the Taylor & Francis Group, an informa business

Typeset in Times New Roman and Futura
by Keystroke, Station Road, Codsall, Wolverhampton
Printed and bound in Great Britain
by TJ International Ltd, Padstow, Cornwall

British Library Cataloguing in Publication Data
A catalogue record for this book is available from the British Library

Library of Congress Cataloguing in Publication Data
Parker, Simon, 1964-
Cities, politics, and power / by Simon Parker.
p. cm.
Includes bibliographical references and index.
1. Cities and towns. 2. Municipal government. 3. Sociology, Urban. I. Title.
HT151.P345 2010
320.8'5—dc22
2010013449

ISBN: 978–0–415–36579–6 (hbk)
ISBN: 978–0–415–36580–2 (pbk)
ISBN: 978–0–203–01828–6 (ebk)

To my family

Contents

Figures

Acknowledgements

I am grateful to Malcolm Miles and John Rennie Short for inviting me to contribute this volume to their Critical Introductions to Urbanism and the City series, to Andrew Mould, Michael Jones and Faye Leerink at Routledge for their continuing encouragement and advice, and to the publisher's anonymous reviewers who provided invaluable comments and suggestions on the original manuscript. A proportion of this book was researched and written during my time as Visiting Associate Professor in Sociology at the University of British Columbia, Vancouver. I would particularly like to thank the Head of Department, Professor Neil Guppy, and Professor Thomas Kemple for the hospitality they showed me, and the students and staff of the Department of Sociology at UBC for making my time there so intellectually rewarding. Colleagues and graduate students in the Departments of Politics and Sociology at the University of York have allowed me to share and exchange ideas on many of the themes that I have developed in this volume. I owe a particular debt of gratitude to Professor Roger Burrows for inviting me to participate in the ESRC e-society research programme on geodemographics and social sorting, some of the results of which feature in Chapter 8 of this volume. Various chapters in this book began life as conference and seminar papers and I would like to thank the participants in the Boston meeting of the American Association of Geographers panel on 'new directions in urban theory', the Departments of Geography at the University of Wisconsin, Madison and at York University, Toronto, and especially Bill Sites of the School of Social Service Administration at the University of Chicago.

Gordon Macleod, Rowland Atkinson and Mike Savage provided expert advice and suggestions on the chapters relating to urban governance, landscapes of power and the politics of identity for which I am very grateful. It goes without saying that any remaining flaws and deficiencies are entirely my own responsibility.

John Foot, Will Montgomery and Nicky Busch have been welcoming and generous hosts during my numerous research visits to London.

Finally, I wish to thank my wife Esmé, and my daughters May Beth and Lily who have been a constant source of inspiration and support to me, and more forgiving of my frequent and lengthy absences than I had any right to expect. It is only fitting therefore that this volume should be dedicated to them.

Part I
Introduction

1 Making sense of urban power

The aim of this book, as the title suggests, is to explore the relationship between cities, politics and power with the hope thereby of shedding light on how cities have contributed to the thought and practice of politics, how cities can be considered both an arena and a site of power, and why the study of the urban complex can help us to rethink and revise the ways we have traditionally thought about political power. But how to make sense of the relationship between such concrete, particular and abstract entities? Such an analysis requires a critical examination of the forms and processes of political power in operation at the urban level and how urban politics operates and interacts with lesser and greater scales of government and networks of power.

Politics, however, is not just a process-based or relational phenomenon; it is also situated within the urban lifeworld in often-unremarked ways (the maximum elevation of buildings, the location of pedestrian crossings, the maintenance of sewage systems, for example). We therefore need to explore the ways in which 'the political' is inscribed within the urban landscape and manifested at the level of the subject (individual, group, social). This requires us to look at how identity and representation are mediated through symbolic, ideological and informational strategies that both create opportunities for and impose limits on political life.

Thinking about the city as both an agent and a site of power also requires us to challenge some conventional understandings of what makes the urban, how the urban is distinguished from the 'non-urban', and even perhaps to reconsider the value – in social scientific terms at least – of thinking about the city as a bounded territorial entity to which defined populations, economic activities, institutional bodies, means of communication and cultural values correspond. It would be easy to assume that the increasing interconnectedness of nations, regions and cities has de-territorialised and de-localised the nature of politics and the organisation of politics and power in the city – leaving us with only the historical ruins of once proud civic democracies. Such a view requires qualifying, first because historically self-contained urban polities have tended to be exceptional and short lived, and second because while the agents of globalisation may see the world from the comfort of the club-class cabin, the vast majority of the world's population, as Jenny Robinson reminds us, live in ordinary cities where 'on the ground' politics continue to matter a great deal (Robinson 2006).

The problem of urban power

With such a plethora of books and articles dealing with urban themes ranging from new regionalism to globalisation and neo-liberalism to gentrification it would seem that the question of urban power is not so much an absent as a ubiquitous feature of contemporary urban studies, and in particular the sub-fields of political geography, urban policy and politics, urban political economy, urban political sociology, urban anthropology, migration studies and so on. However, amidst all this diversity and intellectual effort it is surprisingly difficult to find articles or chapters, let alone monographs or edited collections, that deal with the question of urban power in its own right.

It was not always so. Back in the 1950s, Gerth and Wright Mills devoted the whole of the second part of Max Weber's collected essays to 'Power' within which discrete essays on 'structures of power', 'bureaucracy', 'charismatic authority' and 'the meaning of discipline' were to be found. Browsing through contemporary scholarly articles that claim to address some aspect of 'urban politics' one quickly abandons any hope of encountering even as broad a conceptualisation of power as that contained in Weber's short essay on 'class, status and party'. This is not intended as a criticism of the intellectual credentials of these publications so much as a phenomenological observation: the study of power in the urban–metropolitan–regional–substate context has become divided into a series of discrete research fields each of which has a very good hold on a particular aspect of urban power but where a more synoptic view is palpably lacking.

The first task that confronts us when we take on the difficult subject of researching power in a particular social context is to frame the question. The political scientist would probably still begin with the set of questions asked by Harold Lasswell in 1936, 'Who gets what, when, and how?' The economic geographer, 'Who works where and who makes and buys what and how?' The human geographer, 'Who lives where, when and how?' The sociologist, 'Who gets born, educated, married and dies where, when and how?' The criminologist, 'Who gets arrested or assaulted when, where and how?', and so on. Of course it is perfectly possible to answer such questions empirically and descriptively, as the vast majority of studies do, without considering the problem of 'urban power' in the round. But nevertheless the 'why question' remains a consistent and unavoidable one for those who aspire to dig below the surface of the city's phenomenality.

Politics is not just a question of resource distribution; it is also about how various forms of power constitute political resources in their own right. The personnel that constitute the decision-making and taking bodies – the who? in Robert Dahl's quest for the sources of urban political power – are relevant to this investigation (Dahl 1961), but according to Max Weber, the more pertinent question is what governs? Or rather, what ensemble of social, economic and political forces constitutes the government of the state? Of the three types of power identified by Weber – economic, social and political – it was not as clear as it was for Marx that the former is more potent in the determination of human relations than the latter two.

Class and status can be just as empowering or restricting as the occupation of administrative or political office, and of course in the real world such categories inevitably overlap. Michael Mann draws a distinction between the state's deployment of what he calls despotic or forceful power and that of infrastructural power, which sustains and regulates civil society (cited in Mörner 1993: 3). Because cities are centres of highly concentrated state infrastructure, most of the work of urban policy makers is dedicated to managing, maintaining and extending the machinery of the local and regional state. However, because there is a tendency to regard the provision and allocation of public goods in cities as essentially a techno-bureaucratic activity, it is easy to lose sight of the fact that the power relationship between the (local) state and civil society remains disproportionately in favour of the former (Pahl 1970).

When we ask why cities have the social, economic, cultural and spatial characteristics that they possess and why there appears to be such huge variations in levels of income, segregation, human security, environmental quality and so forth, we must inevitably address aspects of the constitution of urban power. Scholars have sought to map and analyse the changing urban landscape in a variety of ways, the most well-established of which is the field of gentrification, closely allied to which is work on spatial segregation (principally though not exclusively in terms of 'race' or ethnicity). The second looks at the city as an agent of economic transformation that we could loosely group under the heading of 'urban political economy', which also includes work on the urban dimensions of globalisation and neo-liberalism. The third relates to the settlement and growth of cities, especially in relation to regional and international migratory flows and population displacements. The fourth deals with issues of inequality in terms of access to labour markets, types of occupation, remuneration, housing, health, education, utilities, amenities, culture and quality of environment. The fifth is concerned with the city as an arena of public policy and includes the study of different levels of government intervention and includes the organisations, pressures groups and parties involved in the policy process. The sixth is distinct from the last category in that it is interested principally in urban social movements and groups that are challenging of the status quo ranging from violent, hierarchical, paramilitary organisations to non-violent, loosely connected, alternative communities.

This list is not intended to be exhaustive and there are of course overlaps between the different research clusters that I have identified. What is generally lacking, however, is a critical treatment of the city that moves beyond the synchronic particularity of local power processes – i.e. 'power in cities' – to a fuller consideration of the systemic and structured means by which the possibility of urban society is maintained – i.e. the 'power of cities'. Traditionally, much of the attention of urban researchers with an interest in urban power has centred on the institutions of government, and specifically the institutions of democratic control and organisation and the civil and political parties and interest groups that surround them. In recent years, however, there has been a shift of emphasis away from the study of formal (i.e. legitimate) authority towards a more open notion of political power that in Martin Bang's words is 'not to be identified by the terrain of the modern centralised state' but which is indicative of '[a] more interactive, negotiable, dialogical and facilitative authority' which is needed to help people govern themselves (Bang 2003: 8).

As the humanities and certain social science disciplines have become increasingly hostile to notions of hierarchy, essentialism, a priorism, positivism and naturalism, so a small but growing number of political scientists has begun to study and write about political society, not as a primordial and timeless inheritance, but as socially constructed and mediated, and where 'knowledge and power are non-hierarchically intertwined qualities that emerge out of a recursive and interactive covering inside a given terrain, field, system or community' (ibid.). This is not to deny that the business of politics continues to be conducted in and through institutions, which in the case of cities can often be identified in terms of different sites of government, legislative assemblies, judicial and legal apparatuses and political organisations with an interest in participating directly or indirectly in the decision-making process. Rather the argument of this book is that institutional politics is but one aspect of the complex of urban power, and that to understand what role political institutions contribute to the life of cities we have to be alive to the interactions between different terrains, fields, systems and communities of power as suggested above.

One advantage of the institutional approach is that the boundaries of the political are necessarily well defined (Peters 2004). Institutions are above all rule-governed bodies with formal systems for the recruitment of members and for the organisation of their internal and external affairs. Such rules are largely, if not always, transparent and subject to scrutiny – as are their processes of deliberation and decision-making. It may often be the case that the formal mechanisms of power to which institutions are expected to conform differ from those that actually exist, but at least the researcher can begin with a series of benchmarks against which to measure the deviation from formally established norms. This is what Max Weber understood by 'legitimate authority' (or domination) to which he contrasted the non-rule bound world of 'non legitimate authority' (*nichtlegitime Herrschaft*) found in the western medieval city.[1] This departure from traditional domination based on princely authority had its antecedents in classical society but only found full expression through the political articulation of popular collective agents such as the demos, the plebs, the commune, the popolo and so forth (Swedberg and Agevall 2005: 64–5).

Approaching *Cities, Politics and Power*

Given that authority is but one instance of power, how might we think about the realm of 'the political' in relation to the other forms of power that are articulated in and by cities? In attempting to survey the territories of urban power in the pages that follow I hope at least to signpost how some of these broader understandings of the political within an urban context might be tackled.

The volume as a whole is divided into five parts including this introduction and a conclusion. The second part of this volume explores how the 'double helix' of cooperative self-government and violent, coercive domination is intertwined in the history of urban development. Chapter 2 shows how the growth of urban society contributed to the formation of particular forms of organised power structures that differentiated towns and cities from rural societies. As these political formations assumed the more distinct forms of the social movements, organised interests and political parties that we recognise today, the nature of urban government changed from a broadly self-serving and self-selecting oligarchy to an increasingly legal–rational state that began to assume a diverse range of civic and economic functions.

It would be a mistake, however, to view the civic city as a natural and inevitable outcome of the 'democratic evolution of cities', and in Chapter 3 our focus turns to the uncivil city, which over the course of history has characterised much of the urban world, and which persists as the undemocratic, illiberal and intolerant face of many urban societies today. Indeed violence and conflict continues to be used as a key resource of power and domination in the control of cities in ever more terrible and ingenious ways. Examples drawn from cities in conflict highlight the ways in which both legitimate and non-legitimate force has reduced and is reducing the right to the city around the world. At the same time there have been many instances of grass-roots contestation that have changed the course not just of the history of a particular city, but very often of the nation or even the continent of which that city forms a part.

This theme is more fully explored in the opening chapter of the part dedicated to urban governance where the emphasis is chiefly on the modern period (from the closing decades of the nineteenth century to the opening decade of the twenty-first century). In Chapter 4 we examine the key roles played by political parties, organised interests and urban movements in the political life of cities. Political movements have long coincided with the development of the metropolis, and at times particular cities have come to symbolise a revolutionary combination of cultural, social and political attitudes and values, while at others they have acted as mediators between collective interests and the local and national state. The development of urban politics in the United Kingdom, Canada and the United States provides a counterpoint to the emergence of grass-roots community-based urban movements in the Philippines and Brazil where political institutions are more 'porous' and urban civil society less regulated than in the metropoles of the Global North.

Chapter 5 seeks to explore the different modes of governance within and between cities by moving beyond the narrow framework of the question posed by Robert Dahl – 'Who governs the city?' – to the broader and more challenging question, 'How are cities governed?' The theories and approaches that have been developed by Dahl and his contemporaries in order to explain the workings of urban government in liberal democracies are explored and reviewed along with more critical perspectives that emphasise the structural determinants of urban power. The chapter then goes on to explore the varieties of governmental organisation by contrasting the operation of a complex globalised urban polity such as London with the rapidly evolving urban system that is the result of very different state–civil society traditions in mainland China and Mexico.

In Chapter 6 we explore the structural, scalar and global dimensions of the urban complex by charting the new forms of urbanisation that are associated with an increasingly integrated world economy. The rise of the 'city–region' and its policy response in the form of 'new regionalism' reveals how the familiar hierarchical location of the metropolis within a national–state system is breaking down and giving rise to new urban–regional agglomerations that are defying national and even continental borders. In the shift from a world economy and state system built on the certainties and social guarantees of Fordism and Keynesianism to an increasingly neo-liberal, competitive and 'flexible' regime of accumulation, the reconfiguration of the local and regional state has been accompanied by a global competition for investment, for skilled labour and for a bigger share of the global market in goods and services. If there were any doubt that cities have become global enterprises in their own right one has only to note the ever more lavish and increasingly desperate attempts by would-be venue cities to host future meetings of the Olympiad, the FIFA World Cup or to become a European Capital of Culture. The chapter concludes by asking: what are the implications for the configurations of urban governance of the emergence of 'the competitive city', and can we see a decisive shift away from a regulationist, legal–bureaucratic local state to one that is increasingly market-oriented and market-based?

Part IV of the book moves the discussion to the subjective, reflexive and communicative domains of power where the focus in Chapter 7 centres on forms of identity in the context of the urban lifeworld. In this chapter, identity is understood both as a means of self-affirmation and self-realisation and as a category of ascriptive difference and 'othering'. Here we examine how positive and negative identity formations are asserted and contested in cities through political resource mobilisation, along with other markers of social identity such as consumption habits, cultural engagement, habitation and social interaction, each of whose political consequences and potential we illustrate through appropriate case studies. Chapter 8 deals with the emergence of the information city and looks at the way print, broadcast and digital media are defining and determining how the public experiences the city and its diverse populations and users. Information and media networks concentrate in cities and are increasingly displacing formal political assemblies as the new *agora* of our urbanised polities. This chapter explores the impact of information and communication strategies on the processes and structures of urban power and the ways in which information networks now have the power to change the very organisation of cities themselves. Chapter 9 extends the analysis of the previous chapters to the consideration of urban form and the built environment. Here the focus is on the way in which different power strategies are manifest in the urban landscape through, for example, the routing of an urban motorway, the location of a public housing project, or the siting of luxury apartments. Case study illustrations show how the deployment of new technology is increasingly blurring the distinction between public and private space and further widening the gap between affluent and poor inhabitants in cities around the world. In the final and concluding part, Chapter 10 offers some reflections on how politics and power in the city can be studied and better understood in the context of an increasingly cosmopolitan and globally integrated urban world, and offers a prospectus for future research on urban power.

Inevitably, by dividing the vast field of urban politics and power into a series of chapters one risks applying false divisions and sometimes even perhaps creating odd bedfellows. There are reasonable objections to surveying the growth of urban political organisations in one chapter and the work of urban government in the next, given that in democratic polities, at least, political behaviour is recursively linked to the deliberations of urban policy makers and the operations of the networks of urban governance on the ground. Similarly, one might question whether the entrepreneurial city and the information city merit separate discussion since the latter is largely the consequence of a massive shift in global investment patterns towards mostly urban-based digital economies, while the former is increasingly reliant on both the hardware and software of networked urban infrastructures in order to bring both its goods and its customers to market.

Because power is so transcendent and intrinsic to every facet of the urban experience it must follow that the building of discrete categories of analysis is a somewhat artificial task. In its defence I would simply urge the reader to think of the parts of the book and the chapters contained within them as partial segments of a map that have been arranged for the purposes of intelligibility. To follow one's own preferred route through the maze of urban power the reader will need to move forwards and backwards with the aid of some pointers that are included in the text. At the end of each chapter indications are given of further readings that the reader may wish to consult in order to discover more about the themes discussed.

Part II
The political life of cities

2 The civic city

The emergence of urban societies

> One principle emerges ubiquitously . . . [The] need of cities for a most intricate and close-grained diversity of uses that give each other constant mutual support, both economically and socially.
>
> Jane Jacobs, *The Death and Life of Great American Cities*

Introduction

Jane Jacobs talked of there being essentially two methods of human existence – one is associated with 'taking' and the other with 'making' (Jacobs 1985). If this was true of 'primitive societies' it is no less true of the contemporary world where we can observe what Peter Taylor describes as a persistent division of labour between 'protection and associated work' (police, soldiers, judges, regulators, politicians, etc.) and 'production and associated work' (artisans, traders, factory workers, bankers, managers, investors, etc.) (Taylor 2007: 136). Of course this division of labour was not necessarily a permanent one – peasant farmers have often been recruited as soldiers and factory workers can as easily and more profitably be engaged in the production of weapons and instruments of war as non-lethal commodities. In other words, the sword and the ploughshare are not alternative paths of human development: they are sometimes complementary and at other times conflicting means of achieving and sustaining the 'power over life' that is characteristic of civic communities, which in their most dense and complex form we recognise as cities.

Generally, historians, social scientists and archaeologists have sought to distinguish urban settlements from rural habitats by emphasising the size and density of the population, its internal differentiation, the relationship of the settlement to the surrounding 'hinterland', and in particular the existence of economic and political organisations that serve to regulate both horizontal relations (such as the exchange of goods and services between households) and vertical relations (such as the making and enforcement of laws and customs, the decision to engage in armed conflict, the raising of moneys to pay for military campaigns, or for important public buildings such as temples).

All these characteristics of urban life are essentially concerned with how communities of a certain size and density organise their activities through both consensual and non-consensual means. *Kratos*, the Greek term for power, which provides the suffix for many of the contemporary descriptors of political regimes – from democracy to theocracy and autocracy – can be realised only in the context of the civic city whose ancient foundations pre-date those of Plato's Athens by several thousand years.

The origins and development of the civic community

The distinguished archaeologist and historian V. Gordon Childe established 10 criteria that he argued must be simultaneously present in order for an urban civilisation to exist, and such characteristics need to be contrasted with older or contemporary village communities. The first cities must have occupied a greater continuous territorial area and have been more densely populated than previous settlements. Examples would include the ancient cities of Sumeria and northern India where populations were thought to range from 7,000 to 20,000 inhabitants. Second, the composition and function of urban populations differed from that of rural inhabitants. Urban dwellers continued to practise farming and fishing in large numbers and would have maintained plots for the subsistence of their families in the surrounding countryside, but this population was supplemented by artisans, transport workers, officials and priests, all of whom depended on an exchange economy made possible by surplus agrarian production and the trade in other natural and human-made resources. Third, even very small or marginal surpluses were extracted as tithes and taxes by legitimate territorial authorities (the priesthood on behalf of deities, or a divine king). Fourth, monumental public buildings (most often in the form of a temple) symbolised 'the concentration of a social surplus'. Often temples were attached to public granaries, and to the living quarters and workshops of artisans and masons. Fifth, the beneficiaries of such surpluses tended to be high-status elites (priests, warrior chieftains, members of royal courts) who were not required to engage in manual tasks but instead to maintain the historical records of the community, to issue and enforce laws, and to interpret the natural and spiritual world for the benefit of the general population. Sixth, such urban elites were obliged to introduce systems of recording and to formulate scientific principles that allowed for exact measurement. This was accompanied by the invention of standardised writing and numerical notation that allowed the non-oral and asynchronous storage and transmission of information. Seventh, a further development from the advent of writing (or rather the invention of scripts) was 'the elaboration of exact and predictive sciences' such as arithmetic, geometry and astronomy. These epistemic systems made it possible to devise calendars and to anticipate meteorological and climatic change, which led to improvements in agriculture and also facilitated exploration and navigation essential for the establishment of new trade links and the growth of commerce. Eighth, the existence of agricultural surpluses allowed the support of artistic vocations and to advances in artistic technique and aesthetic expression. This also allowed for the inter-generational transmission of craft and artistic skills and the emergence of distinct and sophisticated aesthetic styles within and between different urban centres. Ninth, surpluses were also used to procure non-locally occurring raw materials required for production or religious worship and these were often acquired over long distances. For example, Mesopotamia regularly traded with the Indus valley. Finally, the city was associated with the specialisation of labour, and craftworkers were accorded special privileges and status based on their residence rather than kinship. The association of vocation with urban residence thus gave rise to the new social subject of 'the citizen' (literally city dweller) and thus a political as well as an economic relationship was established that was to further distinguish the people of the towns from the people of the countryside (Childe et al. 2004, 112–15).

Childe's criteria are very helpful in terms of identifying the development of the city as a distinct socio-spatial phenomenon, but while the existence of agricultural surpluses may have been a necessary condition, it was by no means a sufficient one. We need therefore to think about other possible explanations that help to explain the move from scattered to more concentrated population groupings in the ancient world. Crucial to the process of urban formation for Childe and Mellaart were the advent of more efficient tools that 'led to the cultural advance' (Kohl and Wright 1977: 273). This view is also shared by classical political economists such as Adam Smith who believed that, '[t]he cultivation and improvement of the country [. . .] which affords subsistence, must, necessarily, be prior to the increase of the town, which furnishes only the means of conveniency and luxury'(Smith 1776 [1843], Vol. III: 3). However, there could have been no real expansion of towns if the cultivation of farmland had been limited to supporting only its dependent population. The willingness of urbanites to engage in the more risky enterprise of manufacture and long-distance trade and speculation gave rise to the formation of 'human institutions' (marketplaces, legal systems, currency, etc.), which had they not 'disturbed the natural course of things, the progressive wealth and increase of the towns would, in every political society, be consequential, and in proportion to the improvement and cultivation of the territory or country' (ibid. 6).

The debate over the origins of cities is far from settled, however, with Childe's 'technological determinist' explanations of the 'urban revolution' (influenced particularly by the work of Marx and Engels) vying with more ecologically derived theories that stress the adaption of populations to resource constraints and threats. For example, one reason advanced by Maschner and Ofek (Maschner 2003, Ofek 2001) for the move from hunter-gatherer to village-based agriculture was a consequence of the sacrifice of resource maximisation by smaller family groups for socio-political efficiency, 'which could have a very selective advantage in large population aggregates' (Maschner 2003: 284). This evolutionist argument suggests that from very early in the development of human societies, the optimisation of security and the well-being of the community (see, for example, Yoffee 1995: 284) was placed above the higher crop yields or better forage potential that more isolated farming households might enjoy. In other words, the formation of multiple household settlements offered better long-term survival prospects than the dispersal of population groups.

The oldest urban settlement known to archaeologists, Çatalhöyük ('Çatal' means fork, and 'höyük' mound in Turkish), in southern Anatolia is believed to have been constructed between 7,400 and 6,200 before the Christian era (BCE). Covering a site of some 13 hectares, this double mound on the Konya Plain may have been home to up to 10,000 people who would have been engaged for the most part in some form of arable agriculture or animal rearing (Balter 1998; Hodder 2002: 178). However, current archaeological excavations at Çatalhöyük are beginning to dispute the claim of its original discoverer, James Mellaart, that the settlement constituted a Neolithic town (Mellaart 1967). If cities are to be truly distinguished from rural settlements, according to Childe's criteria, there must exist surplus agricultural production with which to support artisans, priests, merchants and others not directly employed in farming or fishing. Whereas, in the view of one of the more recent archaeologists of Çatalhöyük, Guillermo Algaze, the ancient Anatolian site amounted to little more than a large and densely concentrated farming community, or 'an overgrown village' (Balter 1998). This claim is consistent with an earlier sceptic, Glyn Daniel, who argued that neither 7000 BCE Jericho nor Çatalhöyük could be called civilisations, 'but were large settlements that could be called towns or proto-towns . . . or we might label them just as overgrown peasant villages' (Daniel 1968 in Soja 2000: 29). This view is disputed by Ed Soja, however, who argues that an agricultural-led synekism consistent with urban formations was present in both Neolithic sites along with the 'precession' of urban traits among which are included the construction of common walls around the settlement, extra-mural trading networks and evidence of planned public works (Soja 2000: 29–31).

Whether we consider Çatalhöyük a village, a town or a city, it is clear that in the words of the director of the most recent archaeological investigations of the site, 'we are on the edge of a new type of understanding of a mythical world deeply embedded in a complex social system' (Hodder 1996: 366 in Lewis-Williams 2004: 29). Lewis-Williams sees the unusual design of the buildings (typically access to the main living quarters was gained via a ladder through an opening in the roof) as indicative of a tiered cosmology in which the very organisation of the buildings took one down to 'a complexly constructed nether level of the cosmos that had social implications' (Lewis-Williams 2004: 36). While as Hodder writes, 'much of daily subsistence was carried out at a small-scale level . . . there was undoubtedly cooperation at supra-house levels'. It also seems that the location of the site allowed access to lime and other plastering materials that were elaborately decorated and this material was fired to make high quality floors. Geometric designs were also commonly associated with burial areas of the younger population and the use of statuary and bas reliefs point to a strong relationship between aesthetic production and the commemoration of dead ancestors (Hodder 2002: 177).

Although such organisational characteristics of the settlement cannot be said to constitute a formalised polity (or city-state), the existence of a common belief system, the evidence for cooperative social behaviour, and of a rudimentary division of labour are consistent with the formation of political communities in later urban societies. If Hodder is right to refer to Çatalhöyük as a 'complex society', then if we follow Rothman's definition,

> [c]omplexity describes a process during which a social transformation occurs to a qualitatively and quantitatively different kind of economic, governmental, and religious interdependence among people living in close contact in a multisite society. At the heart of that interdependence is the functional 'segregation' (the amount of differentiation and specialization) and 'centralization' (the degree of linkage between the various subsystems and the highest-order controls in society) of the members of these societies.
>
> (Rothman 2004: 76)

Some of the features that Rothman describes have been found to be present at Çatalhöyük, although the evidence of 'a variety of distinct groups based on dimensions like ethnicity, social rank, gender, occupation, etc.' has not emerged thus far from the archaeological record (Hodder 2002: 178). Mellaart, on the other hand, takes the examples of Çatalhöyük and Haçilar to argue 'that cities should be defined by their ability to generate internal growth and should be regarded as cultural centers for a surrounding territory'. Kohl and Wright go on to note that for Mellaart 'these cultural centers appeared at the very beginning of the Neolithic and were by no means self-sufficient', which would tend to contradict Childe's more linear account of 'the urban revolution' (Kohl and Wright 1977: 273).

The dispute over the characteristics of urbanness in more recent archaeological and historical debates centres more around the existence or absence of 'social complexity' than spatial density. Hence, for Stein 'complex society' is 'a deliberately broad concept meant to subsume empires, states, and those early forms of hierarchically organized polities that are generally (but not without debate . . .) called "chiefdoms"' (Stein 1998: 1).

This linking of complexity to state formation while retrospectively appealing from the standpoint of developed western societies, nevertheless runs the risk of imputing a teleological determinism that equates greater levels of political organisation, authority and hierarchy with more sophisticated urban development. Yoffee is more cautious in making this assertion, but is unequivocal in seeing early Mesopotamia as a fully fledged urban-political society built around recognisable state forms

> In early Mesopotamian states, local group autonomy is affected at every level by changing circumstances in political structures, by forces and goals of production and exchange that are set by rulers, and by the new possibilities and constraints on social mobility that are imposed on actors by their embedment in overarching political systems.
>
> (Yoffee 1995: 282)

There is in this statement seemingly nothing to distinguish state–society relations of fourth-millennium Mesopotamia with their contemporary equivalents, which leaves such an approach open to the charge of functionalism that has dogged so much anthropological research on the 'politics' of pre-historical societies in the past. As we will see in Part IV of this volume, the persistent search for 'relations of power' in ancient urban society risks overlooking the ways in which power is materially located in artefacts, in built forms, in techniques of cultivation and husbandry as well as in the cosmology of sacred symbols and practices that do not easily correspond to modern notions of 'the political'.

As Childe expostulates, monumental architecture is closely associated with the emergence of cities and what made such extra-domestic architecture possible was surplus agrarian production. How the ancients turned from being farmers to builders has remained a persistent puzzle in pre-historic studies, but the emergence of a division of labour has long been recognised by social theorists as the point at which modern society began (Simmel 1910; Durkheim 1984). The earliest divisions of labour in sedentary, pre-classical societies were those resulting from priestly or shamanistic vocations and the need to maintain sacred sites or prepare for religious festivals and commemorations including funerary rites and ancestor worship.

The search for central authority in the form of chiefs, kings or high-priests as necessary conditions for the confirmation of cityness in pre-historic society has tended to short-circuit the discussion on the nature of power in the ancient world. A more decentred notion of governmentality put forward by Bang would argue that, '[p]olitical society comprises all those processes of decision and action that go on within its domain or terrain which frames the play of difference' (Bang 2003: 7–8). According to this view, the absence of a central authority does not imply the absence of politics, but rather the existence of a different type of politics where, for example, cooperative rather than hierarchical behaviour – as in Marx's definition of 'primitive communism' – was the norm. Since it is unlikely that the history of the settlement at Çatalhöyük was unmarked by conflict (either from within the community or from external enemies) then decisions would need to have been taken around collective defence or about the repression of violence from within the community.

However, what Stein and many other archaeologists and ancient historians of this period mean by social complexity, it would seem, is not just the existence of processes of decision-making surrounding 'the play of difference', but the existence of a formalised system of government. By this measure, and taking on board Childe's other criteria, it is the ancient urban settlements of Uruk and Ur which flourished in the Euphrates valley some 3,500 years before the Christian era that according to most archaeologists deserve the epitaph of the world's oldest cities.

Ur was originally a port city on the Persian Gulf in present-day Iraq, while Uruk stood beside a major riverine artery of the Tigris and Euphrates whose watercourses subsequently changed thousands of years later to leave a desertified river bed and the remains of these two ancient urban cultures.[2] In the particular case of Uruk, the major expansion of the city and the construction of its impressive ziggurat temples coincided with a vast production of clay tablets that were found throughout the temple states of ancient Sumeria during the Late Uruk period from 3600 to 3200 BCE. At its height Uruk covered an area of 1 square kilometre and accounted for half the settled population of southern Mesopotamia.

The arrival of literacy and the ability to store and convey information through the use of permanent media (such as fired clay) made possible the administration of agriculture and trade, and opened the way to systems of taxation. As writing became more abstract and less pictorial, so other forms of technology developed such as the plough and wheeled vehicles all of which helped to raise agricultural productions leading to greater surpluses. Indeed the frequent depictions on cylinder seals of large baskets filled with grain offered in sacrifice to the Sumerian deities is testimony that Uruk was a highly complex society at the centre of a flourishing inter-regional trade. Oates observes that Mesopotamia has traditionally been regarded as wholly lacking in raw materials by archaeologists and therefore the 'necessity to organize the long-distance acquisition of such resources was a major factor in the early development there of complex political and economic forms'. She goes on to argue that despite some dissent from this view (e.g. Wright 1981), '[t]he development of extensive exchange networks is seen as important in rising social complexity, and especially in the growth of the bureaucratic controls that were developed to record, regulate and redistribute both local production and goods and raw materials from elsewhere' (Oates 1993: 407).

However, it is estimated that less than 1 per cent of the populations of ancient Egypt and Mesopotamia were literate (Schniedewind 2004: 25), and therefore the technical ability to read and write information was tightly restricted to religious, political and economic elites. Knowledge of literacy in other words was intimately connected with the disposition of power in ancient society. As Friedrich Kittler argues, the emergence of transportable writing surfaces (bamboo and mulberry in China, fired and unfired clay in Mesopotamia, papyrus in the Nile delta) was associated with the

> administration of those great river irrigation systems in which cities and high cultures blossomed . . . Thus the same rivers on which the traffic of slave labour and goods flowed simultaneously carried (on the basis of a calendar or goniometric mathematics) the commands of water allocation and the harvesting of products. The same cities that translated the anthropological schema of head, hand and torso into the architectonic schema of palaces, streets and storehouses needed scripts for the processing transmission and storage of their data.
>
> (Kittler 1996)

In other words, the 'informational city' as Castells (Castells 1989; Castells 2000) has termed it is an archetype associated with the pre-modern as much as late-modern or post-modern cities. Indeed, Soja has even gone so far as to argue that the traditional view of the urban revolution that sees cities arising only from conditions of surplus agricultural production needs to be reversed. In his view it 'was cities that were necessary for the creation of an agricultural surplus' (Soja 2000: 35). Even in certain pre-literate ancient settlements such as Çatalhöyük the existence of representational art points strongly to a society that is capable of reflecting on its own spatiality and relationship to nature.

According to Mellaart (1967) the Çatalhöyük wall painting depicts as many as 75 separate buildings against the backdrop of an active volcano and anticipates by several thousand years the clay tablet plan of the ancient Sumerian city of Nippur that has been dated to 1500 BCE (Mumford 1989: 84–5) and the first bird's eye view urban landscapes of the thirteenth century (Soja 2000: 40). (See Figures 1a and 1b.) However, Stephanie Meece has argued persuasively that Mellaart's original assumption that the twin peaked 'volcano' is in fact a leopard

Figure 1a. A wall painting from the Neolithic Çatalhöyük settlement in present-day Turkey (originally published in James Mellaart, 'Excavations at Çatal Hüyük, third preliminary report', *Anatolian Studies* 14 (1964), 39–120, © British Institute at Ankara)

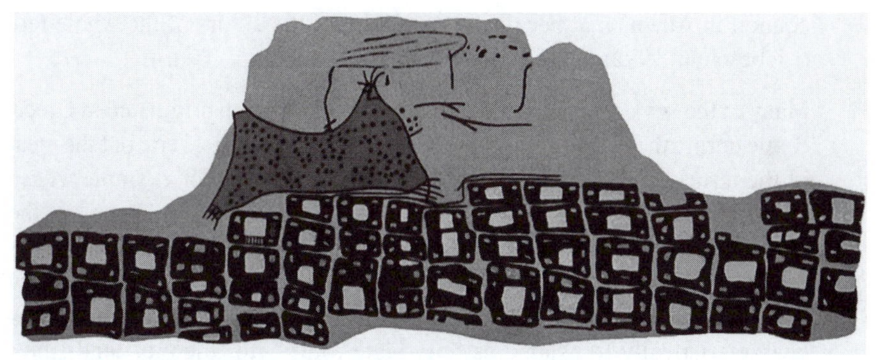

Figure 1b. A drawing of the Çatalhöyük 'map'

skin, while the 'houses' underneath are depictions of geometric decorative tiles that were widely used on the lower parts of the dwellings in Çatalhöyük is more likely (Meece 2006).

It is less relevant to our present discussion whether the ancient Çatal people were the first map makers or volcanologists so much as the clear association of urban settlements with religious devotion (through the construction of shrines), the abundant examples of representational art, and the engagement in trade with other distant communities thus indicating an advanced exchange economy. The capacity of the city to make sense of itself – what we might call the formation of a distinct urban ontology – may not have completely established itself in the Neolithic period, but it was unquestionably present in the eastern Mediterranean regions that today make up Greece, Turkey, southern Italy and the Balkans during the period of classical antiquity.

Forms of authority and political representation in the ancient world

In Moses Finley's celebrated volume on politics in the ancient world, as far as the existence of recognisable political organisation and ideas are concerned, 'the ancient world' encompassed the periods from around the seventh century before and the first two centuries after the advent of the Christian era (CE). But although we recognise the physicality of the ancient cities by their type of architecture and layout, as Liebeschütz argues, 'the origin of the Ancient City was political and administrative', and its 'essential feature was the creation of a political, religious and cultural centre' for a surrounding rural territory. The emergence of Athens as the pre-eminent city of Attica in the classical period depended, according to Thucydides, on the suppression of forms of local political authority across the territory by Theseus and their replacement by a common council in Athens that became responsible for the affairs of the whole territory (Liebeschütz 1992: 1–2).

Many of the key concepts in the western political lexicon originated in Greece and Rome during the period commonly described as the classical era, but the meanings of the terms have changed over the ensuing centuries. For example, Aristotle's understanding of oligarchy was 'the rule of the rich', and democracy 'the rule of the poor', but because then as now the poor have always been more numerous than the wealthy, the latter term has since become associated with 'the rule of the many' (Finley 1983: 1; De Ste Croix 1989: 550). Similarly, the word *polis* is characteristically translated as 'city-state', but both 'the city' and 'the state' existed in changing relation to one another in the classical period, making it impossible to point to regular and consistent features. For example, as Osborne points out, in the democracies of Thasos the country was subordinated to the city, whereas in the case of Elis the villages dealt with the larger town on an equal basis (Osborne 1987: 113–36 in Morris 1991).

We take the features of 'the Athenian model' predominantly from the writings of Plato, Aristotle and Socrates, and these have become in a sense reified as the authentic basis for modern civic life. However, as Morris argues, the *polis* is better thought of as an ideal-type 'complex hierarchical society built around the notion of citizenship' (Morris 1991: 26) than a particular form of government. Lewis Mumford adopted a more demanding set of democratic criteria which for him included 'communal self-government, free communication as between equals, unimpeded access to the common store of knowledge, protection against arbitrary external controls, and a sense of individual moral responsibility for behavior that affects the whole community' (Mumford 1964: 1). As Mumford writes:

> In a great succession of citizens the new urban order, the ideal city, became visible, transcending its archaic outlines, its blind routines, its complacent fixations. For the Greeks added a new component to the city, all but unknown to earlier cultures, dangerous to any system of arbitrary power or secret authority: they brought forth the free citizen.
>
> (Mumford 1989: 160)

Thus the Ancient Greek city provided a model for how rulers and the ruled should behave in a democratic society. Education, reason and the ability to argue were praised in Athenian society as much as the experience of military service. Plato's vision of the Republic demanded a high standard of virtue from its citizens and (contra Mumford) the subordination of individual interests to that of the state. Plato's idealised Republic was one in which the government of the polis comprised philosopher-kings who alone possessed the wisdom to administer justly, and where the public were divided into three classes based on the principle of specialisation. The philosophers provided the top tier and constituted 'the head' of the body politic, the auxiliaries or middling classes provided the heart, and the workers the body and limbs (Plato 1993: xxxiii).

Thus, even in what is regarded as the oldest form of deliberative democracy, power was not evenly shared among the social classes, and order and social harmony were regarded as more important than political equality. The idea that government should be practised only by those with the education and wisdom to rule in the interests of all citizens (women and slaves were notably excluded) emerged from the peculiar circumstances of the Hellenistic polis and the notion that rule by the demos was a necessary condition of statehood had little application outside Ancient Greece during the period of Athenian supremacy.

The ideal city, according to Aristotle, is not to be determined either by the size of its population or the extent of its territory but in 'its common interest in justice' and the common aim of pursuing the good life. Insofar as the ideal city had been perfected in the society and government of the Hellenistic polis (as Aristotle believed it had) then the function of its imperial commanders, such as Alexander, was simply to reproduce, mimetically, the same formal characteristics of Athens wherever its armies triumphed (Mumford 1989: 185) – a technique of colonial urban franchising based on the Hippodamic model that was subsequently imitated by imperial Rome and which was spread throughout the Near East and Byzantium (Demoen 2001: 41). Some, but by no means all of these freedoms, rights and mores were evident in fifth-century BCE Athens, and only intermittently and never simultaneously during any period of imperial Rome, but they were resurrected and given new impetus after the so-called Dark Ages by the emergence of the early medieval city-states that began to develop in Italy, the Low Countries and parts of Germany in the early thirteenth century.

Cities and the emergence of the modern European state

As Liebeschütz writes, '[c]ity organization had been the foundation of the administration of the Roman Empire from the beginning' (Liebeschütz 2006: 309), but with the decline of Rome, the city began to lose its centrality as the privileged site of state power. By the fifth century CE, a variety of central and northern European as well as Asiatic tribes, collectively referred to as the barbarians had begun to occupy much of the territory formerly ruled by the Roman Empire. Although the orthodox historical narrative of 'the sack of Rome' and 'the barbarian invasion' has been slow to shift since the publication of Gibbon's *History of the Decline and Fall of the Roman Empire* in the late eighteenth century, most historians now accept the thesis originally propounded by Walter Goffart that Rome and its neighbouring peoples were assimilated mostly through the peaceful means of legal accommodation and migration rather than by military conquest (though the scale and extent of the migration process is contested) (Goffart 2006).

The extent to which the demise of pre-Christian Rome arrested or disrupted the process of urbanisation in Europe and the Near East is also disputed. Where the writ of Rome had never run, such as in the north-western regions of Europe that now includes most of Scandinavia, as late as the 700s Anglo-Saxon settlements were peripatetic, 'liable to shift gradually within their territorial boundaries or to relocate entirely'. As Smith continues, 'in Denmark settlements continued to wander around in this way until the eleventh or twelfth century' (Smith 2005: 57). However, in the fifth century CE, Goffart argues that there is evidence of the 'prolongation of sophisticated state institutions into the early Middle Ages' along with that of property ownership in the earliest barbarian kingdoms (Goffart 2006: 236). Cities became central to the process of assimilation and accommodation, according to Jean Durliat, because the Roman administration provided the barbarian settlements with a share of the tax revenue that Rome extracted from the landowners rather than sharing the actual ownership of the land itself which would inevitably have led to bitter conflict with the existing landholders (Durliat 1988).

The Holy Roman Empire lacked the unity and cohesion of Ancient Rome – and most crucially the Papacy was forced to rely on the resources of violence that the Emperor was able to co-opt among the local dynasties of the Western Empire. The Eastern Empire centred on Byzantium upheld the Orthodox tradition through a similar alliance with powerful local elites, but the fulcrum of power was to be found within the bishoprics of the Church itself and the courts of the Tsars rather than in the great cities of Constantinople, Moscow or St Petersburg. Much of the impulse towards urbanisation came from what Smith calls 'less egalitarian, more exploitative social relationships, coupled with changing ecological conditions'. In the Italian peninsular, the trend towards the fortification and densification of habitats (known as *incastellamento*) from Tuscany southwards was in part a defensive response to Saracen and Magyar raids, but also weakening princely authority which 'freed landowners to build and garrison their own fortifications, from which they could intimidate their neighbours' (Smith 2005: 57).

The city-state

The government of cities poses particular problems compared to other forms of territory. Unlike the rural peasantry, the typical town dweller in the Middle Ages enjoyed a much greater degree of freedom and where interests were organised such as in the form of merchants' associations, or through the patronage of powerful and rich families – urban government often took the form of a plutocracy rather than an autocracy. The more important cities and towns often also relied on the territorial state to grant special powers and privileges to the urban administration in the form of charters or articles of incorporation. As Friedrichs argues, 'the way that any particular city was governed reflected a combination of arrangements which had been decreed or granted by rulers and agreements that had been negotiated – sometimes peacefully, sometimes violently – by the inhabitants themselves' (Friedrichs 2000: 11).

Governmentality, or what Foucault conceived as the mentality, art and regime of government (Foucault et al. 1991) was a key characteristic of the city-states of Antique and Renaissance Europe where the power to command depended on both symbolic and formal authority. For example, in ancient Roman society, the blessing of the gods was considered far more important than winning the mandate of the Senate, and Romulus and his successor Numa Pompilius were, according to Livy, enthroned according to an augury that expressed the will and satisfaction of the gods (Livy Book 1). There is then – as with ancient Sumerian and Egyptian representations of power and authority – a strong association between spatial and architectonic form, divine purpose and absolute sovereignty.

We also see the idea of the city as an abstract, sometimes sacred, character assuming a personality and role in the administration of power and in the fate of rulers also in medieval Tuscany, whereas Rubinstein (Rubinstein 1958) writes 'the contemporary *specula principum* (the mirror of princes) taught kings to observe the cardinal virtues – the *virtutes politicae* – and such other "political" virtues as were considered essential for good rulership, for instance magnanimity and liberality'. The three cardinal (or political) virtues are Justice, Prudence and Fortitude. The allegory on bad government, or tyranny, depicts the tyrant sitting in the throne of justice with Justice herself prone and bound at his feet surrounded by the vices 'which, like the virtues of the central fresco, have a political significance and do not strictly conform to the traditional schemes: avarice, pride, vainglory; cruelty, treason, fraud; fury, discord and war' (Rubinstein 1958).

Those allegorical virtues that are depicted in Ambrogio Lorenzetti's frescoes in the Palazzo Pubblico in Siena reveal how finely attuned its citizens were to the moral economy of civic order. But the central figure of the prince or ruler as the guardian of all the virtues of statecraft belies the fact that Siena was a republic, governed by a council of elected magistrates whose chamber (the Sala de' Nove) provided the material space for Lorenzetti's paintings. The consensus among historians appears to be that the figure of the ruler is none other than an allegory of the City of Siena itself, for the sovereign is clothed in the colours of the city (black and white) while at his feet lies the Sienese Wolf with the Twins, and above his head are the letters C.S.C.C.V. (or possibly originally C.S.C.V) which either referred to the 'city of the Virgin' (after Siena's patron saint) or the 'city of virtue' (Rubinstein 1958: 181; Skinner 2002: 78–9; Waley and Dean 2010: 124–7). (See Figure 2).

As Rubinstein argues, the strong Aristotelian themes of the allegories could also be read as a panegyric on the relative political stability and prosperity which the merchant oligarchy of the Nove had succeeded in preserving over two generations; and the members of this oligarchy itself could regard the frescoes as expressing a message addressed to the ruling class – that only through unity could they hope to preserve the republican regime and their own ascendancy (Rubinstein 1958: 189).

The Sienese frescoes therefore manifest what Anthony Giddens (1993) has termed 'a double hermeneutic' – in one register they appeal to Aristotelian and Augustinian principles which were assuming the character of a dominant discursive formation in the city-states of Trecento Italy as a legitimation for republicanism and a limited, oligarchic form of representative government, while

Figure 2. Ambrogio Lorenzetti (1285–*c.*1348), *Allegory of Good Government*, 1338–40 (fresco), Palazzo Pubblico, Siena. © Bridgeman Art Library

the lay or doxic characteristics of the paintings are illustrated as a series of distributed benefits as the 'effects of good government' – including the plentiful supply of goods, civil order in the city and countryside, the encouragement of the arts and other noble pursuits.

The 'civic virtues' that are expressed representationally as a legitimation for the unusual form of governance favoured by Siena's magistrates in the thirteenth century are the result of a selective appropriation of classical Greco-Roman and Thomist democratic principles that become cathected onto the persona of the city-state itself. As Fabrizio Nevola writes,

> Lorenzetti's frescoes of the city at peace captured the essence of the relationship between government authority, urban form and social fabric in the pre-modern city. The image was programmatic, laying down an ideal that was to be striven after, and that to a large extent informed the legislative process of urban regulation and change for two centuries following its execution.
>
> (Nevola 2007: 8)

In the following chapter we will see how this celebration of civic republicanism was by no means typical of the forms of authority and the means by which power was organised in Europe and the wider urban world in the period after the decline of the Roman Empire from around the fourth century CE. But the self-governing urban polity was to lay the foundation of what Charles Tilly calls 'consolidated states' (the forerunners of the modern nation-state) in which

> the variable distribution of cities and systems of cities by region and era significantly and independently constrained the multiple paths of state transformation . . . states, as repositories of armed force, grow differently in different environments and . . . the character of the urban networks within such environments systematically affects the path of state transformation.
>
> (Blockmans and Tilly 1994: 6)

Writing later in the Florence of the sixteenth century, Machiavelli was also intent on demonstrating the need for the city's rulers to negotiate the unpredictabilities of fate (expressed in the notions of *virtù* and *fortuna*) by the allegory of a river prone to persistent flooding that must be shored up by armies and dammed by laws and institutions in order to prevent disaster. It is not surprising therefore that political authority should be compared allegorically to the taming of nature since the power of cities has always depended on a nature subjugating technesis – walls, moats and fences to separate humans from nature (and human from human); roads, bridges, viaducts, aqueducts, dams, railways and harbours to pacify, penetrate and transcend nature (for a fuller discussion, see Chapter 9).

Consolidation, urbanisation and the rise of the public sphere

The city-state was peculiarly susceptible to the twin threats of the reassertion of nature against technesis (Pompeii, Lisbon, etc.), rebellion from below, and invasion from outside; to which these sovereign territories responded by developing a largely externally oriented military directed at foreign princes and states, and a police force organised for the purpose of maintaining intra-state order (Bourdieu et al. 1994: 5). Max Weber observed that cities often began life as centres of activity for royal courts, or as army encampments or fortresses, or trading posts that evolved into market settlements (Parker 2004: 10–13). As the size of populations grew the social and economic characteristics of urban inhabitants became more diverse due to the move away from subsistence agriculture and as a result of the increasing complexity of religious and secular organisations. From around the thirteenth century CE the European city was already displaying a consistency of urban form and a spatial distribution of population and activity that distinguished it from urban settlements in the Arabic regions of North Africa and the Middle East or from equivalent population centres in Asia. European urban settlements also became increasingly distinct from their contrary (or *contrata* in Latin, from which the English term 'country' – or the lands and territories that are not occupied by towns – derives) (Epstein 2001: 1). A growing sense of urbanity became apparent in the later Middle Ages as 'urban rights, privileges, and duties were defined and given force by the spaces with which they were associated and in which they established legitimacy' (Howell 2000: 3).

Urban spaces were always at the same time class spaces in that socio-spatial hierarchies were carefully maintained even in the earliest cities. Social status was also very often a function of rights and privileges to engage in particular types of trade or commercial activity, the regulation of which depended on the city authorities or powerful merchant associations such as guilds. As Antony Black writes,

> In Europe the corporate organization of labour and liberal values developed simultaneously and in the same milieu from the twelfth to the seventeenth centuries. Both thrived in towns (to which corporate labour organization was usually confined), and both affected the ideology of the European commune and city-state in specific and discernible ways.
>
> (Black 1984: 237)

Civic bodies, such as guilds, that developed around a particular craft or trade or profession established value systems such as mutuality, solidarity, exchange and equity, which were to be taken up as leitmotifs for a peculiarly European urban civic society – drawing on Aristotle but also inspiring a broader Enlightenment philosophy centred on Idealism and in particular the figure of Hegel (ibid.). As Habermas reminds us, 'Civil society came into existence as the corollary of a depersonalized state authority', which could only supersede the monopoly of legitimate authority in the person and blood-line of the sovereign at the point in which, 'the economic conditions under which [a commodity market] now took place lay outside the confines of a single household' (Habermas 1989: 19). This is consistent with Weber's argument that it was only when urban centres moved from being essentially service and personnel quarters for the royal court to independent market settlements in their own right that the archetype of the modern city could begin to be discerned (Weber 1966).

To be sure, the emergence of 'the bourgeois public sphere' was not a sudden and widespread phenomenon, but its antecedents in the ancient Greek polity and the Italian city-states, and its firm establishment by the seventeenth and eighteenth centuries in the coffee houses of London and the salons of Paris, confirmed the spatial character of civil society as quintessentially urban, and its social character as predominantly bourgeois or *bürgerlich* – terms that both denote the status of city resident and the middling or *Mittelstand* class that began to supplant the first estate of the feudal aristocracy and the second estate of the Church as a dynamic, and at times revolutionary third estate for whom the city became a talisman for progress, culture and enlightenment.

The European Protestant Reformation gave a great impetus to the establishment of 'free cities' such as Calvin's Geneva and the five Swiss city-states including Zwingli's Zürich which declared themselves free from the authority of Rome in the early sixteenth century. Although these 'reformed' cities were more often engaged in fierce theological disputes, which often spilled over into armed conflict, than enlightened plans for a civic commonwealth. So fundamental was the idea of the city as an autonomous community that the political philosopher Althusius proposed that the city-state should be a freely contracting political entity within a wider confederation whose purpose would in part be to limit the sovereignty of greater territorial scales (such as kingdoms and empires). The notion of the city as 'a community of saints' was to become a founding principle for the Pilgrim Fathers – the protestant refugees who established some of the first European settlements in what was to become known as 'New England' in the middle seventeenth century. The lack of hierarchy based on hereditary principles, and a shared belief system that was forged in the conflict against both indigenous and colonial opponents in order to establish political, economic and military dominion, meant that the city and the town assumed a primacy in the American polity that the greatest European urban centres failed to achieve even in the most liberal states well into the nineteenth century (see Chapter 4).

Conclusion

Cities can be understood as arenas, sites, resources, means, nodes, networks and articulations of power. In other words they provide both the form and the content of power routines, systems and modalities as do, on a larger scale, regions, nation-states and international organisations. However, were it sufficient just to regard the urban as a territorial instance of the broader spatiality of the nation-state we would fail to account for the reasons as to why urban settlements (however contested and imprecise the definition) are distinct from other scales and configurations of territoriality and governance, and why cities have retained a consistent social, economic, political and cultural phenomenality over many thousands of years and at the distance of tens of thousands of miles.

One way of helping us to rethink the problem of urban power as 'the organisation of difference in space' is to adopt a methodology whose objective is to identify those features of the urban complex that have persisted, albeit with some variation over time and across space, and which have contributed to maintaining a distinct patterning of governmentality within the formal and organisational confines of a generic urbanism/urbanity. Comparative work on the development of urban societies that aims to identify broad patterns of social and political behaviour has fallen out of favour in recent years because of an increasing reluctance to make generalisations that might attract the post-colonialist charge of 'orientalism' (Said 2003; Said 2004). As a consequence there has been a tendency to emphasise differences rather than similarities between temporal–spatial locales by combining textual with archaeological data in order to view complex societies 'as heterogeneous factionalized entities where culturally specific patterns of ideology, power relations, and competition among socioeconomic groups play key roles in the structuration of polities' (Stein 1998: 4). At the same time, within some fields of social science, an opposition has developed to methods of investigation associated with the natural sciences, and in particular the biological and behavioural sciences because of their basic Darwinian assumption that 'culture' is merely a form of adaptive behaviour to the environmental conditions faced by particular population groups (Aunger 2000).

However, if we wish to understand the reasons why civilisations succeed or fail, as Jared Diamond has argued in a series of landmark books, both environmental and cultural factors need to be taken into account (Diamond and Case 1986; Diamond 2005a; Diamond 2005b). In pre-industrial societies it would seem that a lack of centralised authority can lead to the over-exploitation of the environment or to violent intra-communal conflict resulting in the dispersal and attrition of population. But this is only one in a long list of possible explanatory variables. It is equally possible that over-authoritarian regimes can lead to conflict or to a lack of innovation and creativity and therefore always remain at the back of the pack in terms of economic development and military prowess. In other words, we cannot identify particular forms of political authority or patterns of government and administration with successful processes of urbanisation, but we can point to the centrality of cities for the flourishing of any particular civilisation – whether or not these polities succeed or fail in the long term.

While the notion of an 'urban revolution' has become increasingly challenged by archaeologists and anthropologists in recent years, synekism (or 'interdependence arising from close proximity'), which defined the first city-like settlements, continues to '[manifest] itself as the state through administration based on writing plus monumental building representing the professionalization of ideological, economic, and armed force' (Maisels 1993: 155 in Soja 2000: 25). Where cities flourished as self-governing communities – predominantly in central and western Europe from the thirteenth century onwards – all these factors played a part in helping to create what we shall refer to in the pages that follow as 'material discourses'.

The urban is the space where the materialisation of discourse (understood in its Foucauldian sense as the ensemble of power/knowledge *dispositifs*)[3] reaches its fullest expression. The material discourses that combined to produce the distinctive spatial complex of the civic city can be seen to include ideological ruptures with dominant cosmologies (neo-Platonism in the early Middle Ages, Protestantism in the late fifteenth century), emergent economic categories and classes (money, land and property title, mercantilism, 'free' wage-labour), technologies of communication and control (fired clay tablets, printed books, the census, the telegraph), and socio-cultural praxes (cosmopolitanism, aesthetics, representation, heterodoxy). In the city we find both this necessary contiguity from which shared knowledges are able to develop and reproduce along with systems of regulation and control, both formal and informal, which allow for the emergence of government.

The citizen is a product of what Kant termed the 'unsocial sociability' of men – but while this aspect of synekism can lead towards 'natural government' or self-government, it can also produce a more vicious immanency associated with vice, rebellion and insubordination (Osborne and Rose 1999: 738). Violence and coercion thus constitute the alter ego of the civic city, but the 'uncivil city' has often played a longer and more substantial role in urban history and contemporary urban studies than its progressive, enlightened counterpart, and in the following chapter we try to explore how politics and power operate in these 'cities of dreadful night'.

Further reading

Peter Hall's (Hall 1998) *Cities in Civilisation* remains one of the few contemporary accounts of urban history to rival Lewis Mumford's (Mumford 1989) monumental but still very readable *The City in History*. On the revival of classical ideas of civilisation and the city republics of the late Middle Ages, volume two of Quentin Skinner's *Visions of Politics* (Skinner 2002) is an unparalleled work of scholarship and authority, particularly on the work of Ambrogio Lorenzetti and its context and reception. A shorter essay on Lorenzetti as a political philosopher can also be found in an earlier pamphlet (Skinner 1986). A concise and accessible study of the Italian city-republics is to be found in Waley and Dean (2010).

Charles Tilly remains one of the foremost comparative historical sociologists of urban Europe and both his volume with Blockmans on the emergence of cities and nation-states in the second millennium (Blockmans and Tilly 1994) and his own *Coercion, Capital and European States* (Tilly 1990) are essential reading on this region and period. *The City in Time and Space* by Aidan Southall (Southall 2000) is a fascinating exploration of urban history from an anthropological perspective, with a particular focus on Africa.

3 The uncivil city

Violence, conflict and resistance

> . . . the ubiquity of aggression is an inevitable by-product of living in cities.
>
> Nigel Thrift 2005

> . . . in a town like mine the sign with the classic 'Welcome' is always full of bullet holes because it shows that you are in a place that is under control and that whoever enters needs to know what risks they may be taking . . .[4]
>
> Nanni Balestrini, *Sandokan*

Introduction

In this chapter we examine how violence and conflict have been used as key resources of power and domination in the control of cities. In his monumental account of the development of modern civilisation, *Economy and Society*, Max Weber draws a distinction between what he calls legitimate and non-legitimate authority – the former being associated with what he terms 'the monopoly of the legitimate use of force', and the latter being associated with non-formal types of authority (such as cities) but also other types of association that have the power to coerce and control individuals within a given territory.

In the previous chapter we noted how the emergence of urban civilisation is dependent on the creation of a civic culture in which differences and conflicts between groups and citizens are mediated through the process of law and regulated by a polity, which while it need not be democratically elected and fully representative of all adult citizens, must nevertheless enjoy a sufficient degree of consensual acceptance of its authority for 'good government' to prevail. However, if one studies the actual development of the urban polity over several centuries it becomes clear that rather than there being 'a royal road' from an Athenian-type direct democracy to the pluralist representative democracy of the contemporary western city, a better metaphor would be that of the double helix, where destructive, violent and repressive moments, passages and even entire eras of 'uncivil culture' spiral around the progressive, tolerant and urbane manifestations of the urban complex in an eternal embrace.

However 'legitimate' the monopoly of modern state power may be – whether it takes a subjective (direct) form, a symbolic (impersonal) form, or whether in its systemic form it is constituted by a collective passive acquiescence in the face of dominant ideologies and systems of order and control (Žižek 2008) – violence has always played an important, and often a determining role in the constitution of urban society. The difficulty in constructing an account of the city as both an arena and an agent of violence, conflict and resistance is that these manifestations of negative power occur at so many different levels – from the micro to the meso and the macro to the meta. The effects of urban violence can be found in the emaciated bodies of the poor and the homeless, in the enforced demolition of an urban 'slum' to make way for a new international airport, in the targeted military destruction of civilian infrastructure in Bosnia, Lebanon, Palestine and Iraq, and in the global securities industry simultaneously foreclosing on tens of thousands of homes around the world at the click of a computer key.

When force is used either for the purpose of obliging the victim to do, say or agree to perform some act that they otherwise would not, or to damage or destroy property or goods belonging to another or on which another is reliant it is often with the aim of inducing subjects to act or behave in a way they would not otherwise choose. However, violence can also be an end in itself – it can gratify a lust to see other people suffering or begging for mercy – or even assume the character of a habitual, routinised performance, as when military forces engage in the systematic rape and slaughter of ordinary civilians such as during the Rwandan genocide or in the Dafur region of Sudan. There is nothing intrinsically or necessarily urban about such acts of violence, but in this chapter we explore how the urban context provides a unique set of opportunities and motives for violence that are not found, or are found much less often outside the city.

Places of segregation, spaces of abjection: from the ghetto to the camp

> Fear has shifted from concerns with the physical world and the spiritual realm of salvation during the last four hundred years to the social realm of everyday life. It is other people but not just immigrants – the historical other that have troubled previous immigrants-now-solid-citizens; it is the 'other', that category of trouble that can unseat solid expectations and hopes for a future that can never be realized in what is perceived to be a constantly changing and out-of-

> control world. Fear rests on the borders between expectations and realizations, between hope and reality.
>
> Altheide 2002: 26 cited in Thrift 2005: 139

The Konzentrazionslager Buchenwald (see Figure 3) lies 8 kilometres (or 5 miles) from the charming and historic town of Weimar, whose former residents include Goethe, Schiller and J.S. Bach. Some 56,000 people (mostly men and boys but, in the later stages of the war, women and girls too) passed through Weimar on trains, trucks or on foot to the hilltop camp on the edge of the dense oak forest – never to return. Many of the victims of the Buchenwald concentration camp had been arrested by the Nazis from towns and cities all over Germany and the occupied territories that had fallen under the control of the SS. The great majority were Jews, but there were more than 50 other nationalities present in the camp – among which were political prisoners, prisoners of war, Roma and Sinti, homosexuals, Christians (including Jehovah's Witnesses) and many others considered by the Nazi authorities to be in some way deviant or degenerate.

Outside the main entrance to the camp there is a building, which is often mistaken for the Gestapo political office. In fact the modest single-storey building was constructed after the end of the war, the original having been destroyed either by

Figure 3. KL Buchenwald – Crematorium building. © Simon Parker

bombing or during the evacuation of the camp. But while in residence the state security police were able to use the facilities of the camp's special isolation cells for the purpose of interrogation and torture. They also dealt with the denunciations from the local citizens of Weimar and the surrounding territories, which became so numerous that the Gestapo were unable to investigate them all. Historians have identified over 10,000 such camps – most were established by the Nazis in the occupied territories, and a significant number included ghettos in which the urban Jewish population were forcibly contained, often under the watchful eyes of civilian guards. Daniel Goldhagen identifies 399 such ghettos in Poland, 34 in East Galicia and 16 in Lithuania. Because so many personnel were involved in the administration of the camps and the ghettos – along with the supply of food, materials and logistics – it could be estimated that for every 10 prisoners one member or collaborator with the Nazi regime was necessary in order to sustain this vast trans-continental killing operation. Goldhagen's 'inescapable conclusion is that the number of Germans who contributed to and, more broadly, had knowledge of this regime's fundamental criminality was staggering' (Goldhagen 1996: 169–70).

For the eugenicists (and the Nazis were only the most notorious and successful of the movement's enthusiasts) 'the role of the city was almost accidental – it was merely a space within which natural processes were visible', whereas degenerate traits were a product of inheritance rather than location (Osborne and Rose 1999: 746). The ghetto, for this reason, had to be an urban space beyond which the 'anti-race' of Jews were not permitted to venture, but also a space that was forbidden to other Germans. It provided a physical boundary between the 'pure' and the 'contaminated' whose essential utility lay in its functional separation – whereas, although segregation was also the intention of the original Venetian Ghetto, the designation of this 'othered space' within the Catholic city allowed for the exchange of goods and services that law and custom forbade from within each respective confession (Sennett 2002: 320).

With the advent of modernity we see a passage from 'the ghetto' as a space of segregation and external surveillance to the space of totalising abjection that is 'the camp'. The camp can be seen as the abnegation of the civic city because its inmates are literally stripped bare of their character as human beings – indeed for their tormentors they lack even the status of sacrificial objects. In this condition of 'bare life' (Agamben 1998), law and moral order, these most essential and fundamental aspects of the civic community, are indefinitely suspended. The camp in all its punitive forms functions as a necropolis of the 'social dead' – the deferment of whose physical death is merely a question of expediency or technical–administrative exigency (Goldhagen 1996). Fascism and its nostalgia for a perfect ethno-territorial unity or mono-cultural space can also be seen as a revolt against the cosmopolitan synekism that is the very essence of urbanisation. As Per Binde writes,

> Nazism had its roots in the fiercely antiurban völkisch movement emerging around the turn of the century, in which a fundamental idea was that the Volk (people) constituted an age-old organic unity with the German soil [. . .] Industrialisation and urbanisation cut the bond between man and soil and caused 'uprootedness', the symptoms of which were understood to be moral degeneration, disharmony, and a profound sense of alienation.
>
> (Binde 1999: 177)

In this totalitarian imaginary, the camp becomes a hetero-dystopia for an ever-increasing variety of 'un-German' nations and identities – the stigmatised cosmopolis that the New Order seeks simultaneously to produce and to eliminate. The institution of the Gulag – a space of abjection for 'the enemies of the people' who also needed to be removed as far as possible from the 'healthy body' of Soviet society – also reflects the totalitarian fantasy of control and the purposeful justice of state violence aimed at the decombination of citizens and class enemies or Völk and Untermensch/Unmensch.

The denial of civic or human qualities to the inhabitants of camps may not be as extreme or contemptible in every instance – but it is important to recognise that the same process of 'exceptional statehood' is involved in the establishment and justification of Buchenwald, the Gulag and Guantànamo. In Agamben's phrase the camp is a biopolitical paradigm of the modern in which man's natural life becomes progressively included in 'the mechanisms and calculations of power' (Agamben 1998: 119). Foucault provides an insight into how this discourse of power has changed from the Middle Ages where the abject and despised frequently experienced their punishment in the public gaze – the performative and spectacular sadism of the public execution, as explored most memorably and shockingly in *Je, Pierre Rivière . . .*, eventually gives way to the nineteenth-century occlusion of state homicide behind the walls of prisons, which nevertheless form a forbidding and admonitory presence within the urban landscape (Foucault 1982). In 'post' modernity, the invisibility of state violence becomes an end in itself: the rendering of suspects in unmarked planes from disused airfields to distant interrogation centres and their subsequent imprisonment in the appropriately named Camp X-Ray (whose operation is not subject to the gaze of ordinary civilian eyes) is but the most recent example of the state of exception, which must 'de-civilise' the enemy – and as importantly the state's own norms and standards of conduct – in order to protect through violence and inhumanity, the peaceful and humane values on which its legitimacy and credibility rests.

It is not that the persecution of despised groups and their concentration within physically confined spaces is a peculiar feature of the twentieth and twenty-first centuries, but rather that the technologies of surveillance, sorting and control together with the apparatus of highly integrated mechanisms of violence across increasingly vast geographical scales means that cities become simultaneously protagonists, vehicles for and victims of the organised use of force for political ends.

Targeting the city: urbicide in the age of the 'War on Terror'

Urban populations since the beginning of recorded history have been subject to the organised violence of marauding armies, they have been besieged, burned, captured, enslaved, starved and forced into exile – from the sacking of ancient Troy to modern day Sarajevo. Urban warfare and its successful prosecution, according to Charles Tilly (Tilly 1990), has been central to the success of state building in Europe since the Middle Ages. But, by the eighteenth and nineteenth centuries, battles were predominantly won or lost at sea or in the open country. This was also largely true for the 1914–18 war but, with the advent of the bomber, by the 1930s cities and their urban populations were once again the target of systematic and massacring attacks – captured unforgettably by Picasso's painting of the destruction of the small Basque town of Guernica by German bombers at the time of the Spanish Civil War.

When the US military launched its 'Shock and Awe' attack on Baghdad in March 2003 it was putting into plan a military doctrine for achieving 'rapid dominance' that had been developed by Harlan K. Ullman and James P. Wade at the National Defense University in Washington DC seven years previously. As the authors write:

> Shock and awe are actions that create fears, dangers, and destruction that are incomprehensible to the people at large, specific elements/sectors of the threat society, or the leadership. . . . The B52 raids in Vietnam provided localized elements of Shock and Awe, but until applied to the capital city of Hanoi, had no impact toward war termination. When applied in concentrated repetitive strikes in November/December of 1972 under 'Operation Rolling Thunder III', the cease fire followed in short order. . . . How to apply elements of Shock and Awe against rogue states, terrorist elements, international drug and crime cartels, as well as in the more traditional MRCs [major regional contingencies] and LRCs [lesser regional contingencies] needs much further study and analysis.
>
> (Ullman, Wade et al. 1996: 78)

In other words 'Shock and Awe' is not simply a military tactic but a pervasive form of statecraft aimed at subordinating 'threat societies' through the deployment of spectacular, intense and devastating violence in order to break the resistance of the target population. The emphasis on the need to target densely populated cities such as Hanoi, along with Hiroshima and Nagasaki – described as the 'ultimate military application of Shock and Awe' which brought about the unconditional surrender of Japan – can be explained in terms of the doctrine attributed to General Patton that a good commander does not make his troops die for their country, 'he gets the other SOB to die for his' (Ullman, Wade et al. 1996).

Modern military doctrine does not identify its objective as the deliberate targeting and punishment of civilian populations such as occurred during the 'conventional' bombings of Guernica, Dresden, Coventry and Tokyo in the 1930s and 1940s. Instead what Stephen Graham has identified as 'fetishistic technophiliac fantasy of perfect power'[5] has evolved in terms of urban warfare whereby potential enemies are presented as eminently visible and distinguishable from 'innocent' civilian populations thereby allowing the former to be targeted by smart weapons technology for precise and exclusive destruction. As one senior US air force commander explains, 'the combination of using high-fidelity ISR feeds and guided weapons has given militaries a limited ability to distinguish insurgents from the population and strike them with precision, while mitigating collateral damage' (Brown 2007: 76).

'Collateral damage' is thereby considered a regrettable but necessary human cost that is the moral responsibility of the enemy's cynical 'human shield' ploy of operating where they know non-combatants are present.[6] Since the introduction of long-range high-explosive artillery and bomber aircraft, the technologies of military violence have acquired the power not only to physically annihilate cities without having to fight and conquer them first, but to do so with no regrets.

There are other ways of achieving urbicide, which Graham defines as 'the deliberate denial or killing of the city – the systematic destruction of the modern urban home'.[7] This does not have to involve the complete annihilation of the buildings and infrastructure of the city – which was not even achieved by the atomic bombing of Nagasaki and Hiroshima but by what military commanders refer to as 'the degradation of civilian infrastructure'. This can range from the blockading of roads and communications, to the diversion, abstraction or pollution of water sources, the cutting off of energy and food and medical supplies as well as the physical destruction of homes.

Bosnia

During the Bosnia-Herzegovina Wars of 1992–5, it had become clear that different sides in the conflict were engaging in the deliberate destruction of the urban environment – especially important cultural symbols and internationally significant heritage sites such as the Old Bridge of Mostar (Stari Most) (see Figure 4). So calculating and systematic was this destruction of the built environment that architects began to refer to it as 'urbicide' (Coward 2004: 157; Narang-Suri 2009). Although it had no real military and strategic significance, the Council of Europe reported that the historic core of Mostar 'was clearly targeted by the heaviest guns available to the HVO Bosnian Croat army/paramilitaries', but it was not just architecturally significant non-strategic buildings that were shelled during the conflict, but also 'housing, public institutions, cultural monuments, utility buildings, open spaces' (Coward 2004: 161).

According to Grodach, 'the Croatian effort to capture Mostar was a central component of the secret Karadjordevo Agreement in which Milošević and Croatian president Franz Tudjman colluded to partition Bosnia and annex the new country into Serbia and Croatia, respectively' (Grodach 2002: 67). As a symbol of Ottoman architecture Stari Most 'becomes symbolic of foreign Islamic

Figure 4. Stari Most, Mostar's Old Bridge, after its reconstruction following the shelling of 1993. © Holger Mette/Dreamstime.com

occupation', and its destruction a reassertion of 'European' cultural primacy and hegemony (ibid. 74). As Bollens reminds us, the Bosnian war 'killed over 100,000 people and forced half the country's 4 million people to flee their homes to friendlier locales within the state (1 million people "internally displaced") or to other countries (1 million "refugees")'(Bollens 2008: 196). Hundreds of villages, towns and cities were subject to deliberate 'ethnic cleansing' by ethno-nationalist military groups. According to an indictment of the International Criminal Tribunal for the Former Yugoslavia, the goal of the HV and HVO forces' 'ethnic cleansing' was to gain control of the municipalities of Mostar, Jablanica and others in Bosnia-Herzegovina and to force the Bosnian Muslim population to leave these territories or to substantially reduce and subjugate this population. The means used for this purpose included killings, beatings, torture, evictions, destruction of cultural and religious heritage, looting, deprivation of basic civil and human rights, and mass expulsions, detentions and imprisonments, all of them executed following a systematic pattern of ethnic discrimination. As a result of this campaign, tens of thousands of Bosnian Muslims abandoned Mostar, Jablanica and other municipalities in Bosnia-Herzegovina. The traditional ethnic diversity of these municipalities was virtually eliminated, and an ethnically homogeneous society and institutions were imposed in these areas.[8]

Gaza

Jericho and its surrounding region contain some of the longest established urban settlements in the world. The Jordan valley has also been home to many different ethnic groups who have contributed to a rich and cosmopolitan urban culture that has experienced long histories of peace and prosperity, but also protracted periods of conflict and violence (Fisk 2005). The densely populated and urbanised Gaza strip, which is now home to some 1.4 million people, has been controlled variously by the Philistines, the Pharoic Egyptians, the Babylonians, the Jewish nationalists of Judas Maccabee, the Roman emperors Pompey and Herod, the Knights Templar, the Arabs, the Turkish and Mamelukes, the British, the Egyptians and, for a very brief time during the 1967 war, Yasser Arafat's Palestine Liberation Organisation (PLO – El Fatah), before succumbing – along with the territory of the West Bank – to the subsequent occupation of Israeli forces (Butt 1995).

Gaza was at the centre of the street-based protests against the Israeli occupation which became known as the intifada, and after the Oslo Accords between Israel and the PLO a 'semi-autonomous' Palestinian Authority was created in 1993, which enjoyed limited powers but no control over its borders and coast or links with the other occupied territories in the West Bank. Following Prime Minister

Figure 5. A plume of smoke rises over Gaza City following an Israeli air attack during Operation Cast Lead. © Slidezero/Dreamstime.com

Ariel Sharon's decision to withdraw Israeli forces along with 8,000 settlers in 2005, Gaza and its surrounding territory has become an increasingly bitter battleground between the Israeli Defence Force and Hamas, which in 2007 had displaced Fatah as the sole Palestinian authority in the strip.[9]

Declaring its determination to put an end to Hamas rocket attacks from Gaza on the nearby southern Israeli towns of Sderot and Ashkelon, Israel's Prime Minister Ehud Olmert announced the launch of an air, sea and ground assault on Gaza (Operation Cast Lead – see Figure 5) which lasted from 27 December 2007 to 18 January 2008. As a result of the operation, the Israeli government reported four Israeli fatal casualties in southern Israel, of whom three were civilians and one soldier, killed by rockets and mortars attacks by Palestinian armed groups. A further nine Israeli soldiers were killed during the fighting inside the Gaza strip, of which four were believed to have been caused by so-called friendly fire. Estimates of Palestinian civilian casualties during the Gaza conflict varied from 1,166 cited by the Israeli authorities, to between 1,387 and 1,417 by non-governmental organisations (NGOs) such as Amnesty International, to 1,444 by the Gaza authorities.[10]

The United Nations Human Rights Council, in adopting the report of Justice Goldstone by a majority of 25:6, found that 'while the Israeli Government sought to portray its operations as a response to rocket attacks in the exercise of its right to self defence, the Israeli plan had been directed, at least in part, at the people of Gaza as a whole'.[11] As Stephen Graham writes of an earlier Israeli Defense Forces operation in Gaza and the West Bank (Operation Defensive Shield) the decision to target urban population centres represented a major strategic U-turn for the Israeli armed forces whose doctrine since 1948 was that 'entering cities should be avoided, as this offered no benefits whatsoever. Thus, cities and population centres should be bypassed' (Tamari 2001: 205 in Graham 2004: 205). However, the view of one former Israeli cabinet minister and retired brigadier-general, Efraim Eitam, that 'the weak evasive side addresses the asymmetries of military power by using the building and city as a weapon' (Graham 2004: 206), has persuaded western military elites – particularly in the wake of the events of 11 September 2001 – that urban centres are the new global battleground (Graham 2010). As Saskia Sassen argues, the capacity of cities to triage conflict through commerce and civic activity may be diminishing with the result that new types of violence such as asymmetric war have the effect of 'urbanising war' (Sassen 2008).

The common denominator in all these 'theatres' is the tendency of military commanders and their political masters to regard urban 'threat' societies as subjugatable only by direct force. That this coercive strategy is far from new can be discerned from the comment that '(General) Montgomery and others believed that repression consistently and relentlessly applied could crush the revolt . . . [of the Zionist insurgents]', in 1945–7 as British military commanders believed they had succeeded in doing during the Arab revolt of 1939 (Newsinger 2002: 20 cited in Mattox and Rodgers 2007: 67). The effectiveness of 'dynamic targeting', however, relies on a sophisticated, and up-to-date knowledge of the command-and-control structures of the enemy and the very best human intelligence on the ground – what might be called a biopolitical panoptic gaze.

Comparing the longer-established, decentralised and harder-to-penetrate command structures of Hezbollah in urban Lebanon to the more hierarchical structures of Hamas, according to USAF Major-General Brown, 'Israel's experience shows that, much like treating a cancer, combat operations prove more effective on an immature and isolated insurgency' (Brown 2007: 78).

The use of lethal and destructive force by the state is never simply a means to an end – i.e. 'ending terrorism' – rather it must be understood as Žižek (Žižek 2008) argues as a more spectacular and visible extension of the systemic or routine daily violence which in the case of Gaza, a territory that Avi Shlaim has likened to 'an open-air prison',[12] is realised through the unemployment of half the population, and the dependence of 80 per cent of its inhabitants on less than $2 a day. It is not that such spaces of abjection – the camp, the gulag, the ghetto, the city under occupation – are identical in terms of the bareness of life that they produce, or that all coercive state violence can be reduced to an annihilationist common denominator. Rather the deployment of urbicidal state power needs to be seen as a manifestation of the biopolitical control of territory and space which seeks at the same time to render the violence of the subaltern, the internee and the abject as always and everywhere aberrant – a violence without cause or justification – where often the mere existence of subaltern bodies can provoke and justify the violence of the state (Esposito 2008).

The killing streets: violence and the urban poor

The problem of violence, whether state-sponsored or not, is particularly acute for the poor of the urban settlements in the Global South. Indeed Moser and Mcilwaine argue that increasing levels of urbanisation in the developing world encourage violence where this is associated with high levels of poverty and inequality. Even where states are embarking on a process of democratisation, 'everyday' violence continues unchecked in many poor communities, while the grip of drug cartels and powerful criminal organisations from Mexico and Colombia to Jamaica and China appears to be strengthening. The negative consequences of violence are not only limited to the immediate victims, but economic growth and investment suffers as does the capacity of governments to tackle the systemic violence of poverty, inequality and social exclusion (Moser and Mcilwaine 1999: 203). For much of the Global South, as Anil Bhatti writes, '[c]onflict and violence have become an integral part of the city . . . Instead of becoming an urban space of liberation, the post-colonial city is the locus of disaster' (Bhatti 2006).

Colombia

Colombia is the fourth most populous country in Latin America, and one of the most violent countries in the world. The District Capital of Bogotá, established by the 1991 Constitution, lies 2,630 metres above sea level and is home to around 7 million people. The roots of contemporary urban violence in Bogotá can be traced back to the assassination of the reformist Liberal leader Jorge Eliécer Gaitán in 1948, leading to an insurrection that began in Bogotá and which quickly spread throughout Colombia. In the 10 years that followed 'El Bogotázo', la Violencia claimed the lives of some 200,000 Colombians and over 2 million rural peasants were displaced or forced from their lands (Moser and Mcilwaine 2004: 41–2).

That Gaitán's murder was not a casual act of violence can be discerned from the increasingly vituperative campaign directed against the popular leader known as 'il Caudillo' by the Conservative establishment. One Conservative newspaper openly called for Gaitán's elimination and his personal staff were extremely concerned about the Bogotá police and its Conservative police chief who systematically filled the ranks of the city's force with recruits from the staunchly Conservative regions of Santander (Braun 2003: 132–3). It little mattered that the assassin was apprehended by a passing police officer within seconds after the shooting, or that the murderer was not politically connected, for the incensed crowd it mattered only that the chronicle of Gaitán's death had already been foretold by his establishment opponents in the two ruling parties.

Urban violence thus became a powerful resource both for the government and opposition groups in the following decades in order to entrench vested privileges – especially land ownership – to control urban labour markets and prevent labour organising, but also for the poor to protest at political and economic injustice, and for paramilitary and insurgent groups to achieve dominance within the towns and cities through which the increasingly lucrative drugs trade is routed and distributed.

Many Colombians who are forced into informal settlements surrounding the larger cities are internally displaced persons (IDPs) – the majority of them peasants who have fled conflict zones due to threats from paramilitaries, guerrillas or common criminals (especially those connected with the drug trade). The Colombian government had registered 1.7 million displaced persons between 1995 and 2005 and the United Nations High Commissioner for Refugees (UNHCR) estimated that more than 2.5 million have been displaced from their homes at some point between 1990 and 2005. Because paramilitary and guerrilla organisations such as the FARC and ELN try to discourage IDPs from officially registering by the use of threat of violence, the official figures are likely to be well below the actual numbers of migrants fleeing conflict.[13]

The quadrupling of the murder rate between the mid-1980s and the mid-1990s can be attributed to the intensification of what has been called Colombia's 'triangular' or 'dirty' war fought between the national security forces, para-military guerrillas and powerful drugs cartels – all of which contributed to increasingly high levels of violence that were particularly concentrated in the capital's poorest neighbourhoods and informal settlements (Giraldo 1996). According to the United Nations Settlement Programme (HABITAT), the slum population of Bogotá continued to increase during this period – mostly as a result of rural–urban migration. For this marginal and vulnerable population, physical and social isolation is a serious problem due to the lack of transport infrastructure and persistently high levels of violence that are not captured by the official statistics. The inhabitants of the 'official city' tend to view the slum dwellers as unwelcome and undesirable additions to the city's populations, referring to them as *desechable* (disposable), *gamin* (street boy) or *vagabundo* (tramp) – terms that are often associated with criminality and delinquency (UN HABITAT 2003: 205).

The award by UNESCO of 'City of Peace' came after a decade of conflict when in the mid-1990s Bogotá was the most violent city in Latin America with an official homicide rate of 80 per 100,000 inhabitants. But by 2006 the murder rate had dropped to around 18 per 100,000. The head of the City Council's sub-office for security, Hugo Acero, attributes this success in large measure to the constitutional reform of 1991 that put responsibility for security policy in the hands of governors and city mayors, thereby allowing them to give direct orders to the commanders of the National Police. Better intelligence and information gathering and sharing processes, such as a 'Plan for Security and Coexistence' (which aimed at strengthening citizenship and peacefully resolving domestic, interpersonal and community conflicts ??) has also contributed to the reduction in violent crime. Other initiatives included more specially administrative staff in the area of criminal justice, the introduction of local 'security councils' for each district of the city, external evaluation using private and third sector stakeholders, and citizen participation and training aimed particularly at local community leaders.[14]

In contrast, the homicide rate in Colombia's second largest city, Medillín, was higher than Bogotá's for most of the 1990s and 2000s. This was due, in the view of Sanín and Jaramillo, not so much to the absence of the state (both government and regular capitalist organisations are nearly as extensive in Medillín as in Bogotá), but to the porousness of the state, which instead of ensuring a strict monopoly over the means of violence permits 'competitive, parasitic and mutualist relations with other organizations that command means of coercion' (Sanín and Jaramillo 2004). The decision of the national government to allow the formation of paramilitary groups in 1968 (Law 48) known as 'self-defence units' (Giraldo 1996: 67–8), which operate as clandestine 'auxiliaries to the state and the police' (often with the same personnel), gave rise to alliances with the powerful drug cartels (Livingstone 2003: 78–9) and to a violent competition with insurgent groups for control of the city. As a result 'national or municipal governments' have reached settlements, 'with non-state armed agencies [that] have been characterized by . . . "paradoxical pactism"' – that is to say, 'temporary agreements that solve specific problems, but that do not address the general balance of power that underlies these problems' (Sanín and Jaramillo 2004: 19). State authorities that adopt such a strategy are prone to break down under external pressure and, as Tarrow points out in the case of the Italian city-states, such a breakdown can easily spill over 'into non-institutional contention' (ibid.).

Organised crime and the control of cities

In few parts of the developed world is this 'non-institutional contention' more well established than in the southern regions of Italy itself, where the civic values of the Italian city-state were never really present (Putnam, with Nanetti and Leonardi 1993). Instead in Sicily, Puglia, Calabria and Campania, the political and economic fabric of the urban and regional system has been considerably penetrated by organised crime – at times provoking the deployment of the army on the streets in an attempt to reassert the monopoly of state authority over an urban society that has often borne the cost of government while experiencing little of its benefits.[15]

The United Nations defines an organised criminal group as

> a structured group of three or more persons, existing for a period of time and acting in concert with the aim of committing one or more serious crimes or offences . . . in order to obtain, directly or indirectly, a financial or other material benefit.

However a 'structured group' is defined as one 'that is not randomly formed for the immediate commission of an offence and that does not need to have formally defined roles for its members, continuity of its membership or a developed structure' (United Nations 2000). This apparently contradictory attempt to define organised criminal associations reflects the elusive nature of a form of powerful territorialised criminality that is at the same time non-formal, discontinuous and unstructured. This definition also fails to communicate the fact that organised crime is also one of the richest and fastest growing global businesses, and the city and its hinterlands provides both a lucrative market and a constant source of existing and future personnel.

According to the author of one of the best-selling accounts of organised crime in the Naples region of Italy, the Camorra (or 'the System' as it is known locally) is now 'the most solid criminal organization in Europe in terms of membership' (Saviano 2007: 45). The criminal clans that operate within the province of Naples are genuinely global enterprises with operations in China, Eastern Europe and South and North America. They not only control the lucrative trade in narcotics, for which they have become wholesale suppliers to the Sicilian Cosa Nostra and the 'Ndrangheta of Puglia, but also the importation and export of genuine and fake consumer goods (especially designer wear), construction and the illegal dumping of toxic waste. Even as far back as the 1860s, Pasquale Villari in one of his *Lettere Napoletane* observed in a letter to a Turin newspaper that a young writer of his acquaintance who wanted to write on the subject of the Camorra (perhaps the Saviano of his day) had discovered that all the newspapers of Naples are forced to pay a tax to the Camorra in order to be able to have their editions sold publicly on the streets. Villari also provided an eyewitness account of how two local well dressed *camorristi* would stand at a crossroads in front of the busy Ponte di Casanova with a large basket into which a considerable quantity of fruit and vegetables had been placed by the peasants on their way to market. 'No one neglected his duty', commented Villari, and despite the fact that one of the *camorristi* brandished a large stick, 'the basket was filled with scrupulous care and with apparent goodwill' (cited in Consiglio and Musella 2005: 57–8).

A hundred years later, '50 percent of the shops in Naples alone are actually owned by the Camorra' (Saviano 2007: 50). Meanwhile the sticks have been replaced by guns and explosives, and with deadly consequences. Since 1979 the Camorra clans have been responsible for some 3,600 deaths in Italy – that is more murders than the Sicilian Mafia, the 'Ndrangheta, the Russian Mafia, the Albanian criminal families, and more than those killed by ETA in Spain, the IRA or the Italian Red Brigades and Armed Revolutionary Nuclei, or the total number of deaths resulting from the violent actions of the Italian state (Saviano 2007: 120).

Although it is tempting to think of organised criminals such as the Camorra as 'violent entrepreneurs', as Varese observes, this label 'would be particularly misleading', because

> the camorristi are, first and foremost, individuals specialising in the use of violence. They do strike deals with genuine entrepreneurs, but a fundamental difference obtains between the dressmaking workshop owner and his employees on the one hand, and the camorra boss and his underlings on the other: clans are ultimately armed forces.
>
> (Varese 2006: 271)

Ultimately, violent urban gangs such as the Camorra are able to control a black economy worth hundreds of millions of dollars and to extend their criminal network on a global scale because the state is either unable or unwilling to enforce its monopoly of the legitimate use of force. The autonomy of the local state is not a theoretical debate for the *camorristi*, but a political opportunity structure that allows for complete control of the local territory and further opportunities for the ransacking and distribution of both public and private resources by what Jason Pine refers to as the Camorra's 'nomadic war machine' (Pine 2008: 203). Occasionally, when the political circumstances allow, the legitimate state pushes back by passing legislation permitting the Interior Ministry to dissolve local authorities that have become infiltrated by organised crime. Since the law took effect, in the province of Caserta alone,16 municipalities have been dissolved – five of them twice (Saviano 2007: 231).

But the awkward truth remains that, unlike their Sicilian counterparts, 'the Camorra clans don't need politicians; it's the politicians who need the System'. In Campania it is perfectly possible for left-of-centre parties to hold power at municipal, provincial and regional levels but this is because the local state is often nothing but an empty shell – the key power, the commercial enterprises, are under the control of the clans and this 'allows them to control everything else' (Saviano 2007: 47).

Conclusion

According to the Strategic Studies Institute:[16]

> a new kind of war is brewing in the global security arena. It involves youthful gangs that make up for their lack of raw conventional power in two ways. First, they rely on their 'street smarts,' and generally use coercion, corruption, and co-optation to achieve their ends. Second, more mature gangs (i.e., third generation gangs) also rely on loose alliances with organized criminals and drug traffickers to gain additional resources, expand geographical parameters, and attain larger market shares.

What this report fails to mention is that this 'new kind of war' against gangs, organised crime, drug-traffickers and insurgents is being fought by a state that in many parts of the world has been compromised and corrupted by the same elements against which it is supposed to be engaged. There is a war being played out at the level of the urban habitus, but it is not a conflict between the civil state and uncivil insurgency and criminality. This is a battle for the defence and preservation of the *Rechtsstaat* itself – besieged from one side by the will to power of 'the state of exception' (Agamben 2005) and on the other by a delinquent shadow state, complete with its own security apparatus, bureaucracy, financial system and diplomatic representation.

This chapter has tried to illustrate that power in the form of violence and domination has been as much a part of the urban experience as the law governed civic polity, which since Plato's time has existed as a non-violent ideal community that posits the realm of intellect, law and 'all that belongs to the soul' against the perversion of sovereignty that is Hobbes's *ius contra omnes* (Agamben 1998: 34–5) As we discussed in Chapter 2, the civic community did not arise spontaneously or accidentally in the evolution of cities; it required the mobilisation and sustained organisation of collective interests and intellectuals capable of envisioning a new urban society. In Part III we examine how political parties, interests and social movements have combined to shape the modern urban polity through the emergence of an increasingly industrialised society where cities have become the locus for a highly integrated regional, national and global system organised in and through nation-states that largely enjoy a monopoly of the legitimate use of force.

Further reading

Because the literature on the history of urban conflict and violence is so considerable, a good place to look for signposts is the article by Ailsa Winton, 'Urban Violence: A guide to the literature' (Winton 2004). Jane Schneider and Ida Susser's *Wounded Cities: Destruction and reconstruction in a globalized world* (Schneider and Susser 2004) offers accounts of different experiences of urban conflict and violence from Baghdad to Belfast to Bangkok.

A compelling series of essays on cities, terror and 'urbicide' is to be found in the volume edited by Stephen Graham (Graham 2004), whose *Cities Under Siege: The new military urbanism* (Graham 2010) is a chilling account of how the modern technologies of warfare and asymmetric conflict have put millions of the world's urban populations on the front line.

There are few comparative studies of organised crime and international criminal networks that have a distinctively urban focus, but *Understanding Organized Crime* by Stephen L. Mallory provides a detailed historical study of a number of city-based organised crime syndicates in Colombia, Mexico and the United States and offers a critique of social science explanations for the persistence of organised crime in contemporary society (Mallory 2007).

Part III
Urban governance

4 Political organisations and the quest for urban power

> There are two ways of making politics one's vocation: Either one lives 'for' politics or one lives 'off' politics. By no means is this contrast an exclusive one.
>
> Max Weber, *Politics as a Vocation*

Introduction

In the previous two chapters we considered how urban civic cultures have evolved over time and also why cities have been and continue to be afflicted by violence and non-legitimate power. In the first chapter of this part dedicated to urban governance the focus moves to how and why social actors within cities organise themselves politically, and what motivates people to come together in order to change the way their society is organised – whether at an exclusively urban level or more generally. How do different social, economic, religious and cultural interests manifest themselves at the urban level and through what strategies do they seek to influence the governing process? To what extent is political participation in cities a means to an end (for example, electing political representatives to government office) or an end in itself (for example, taking part in local community organisations)? Although we will examine the government of cities in Chapter 5, as Max Weber showed, the relationship between rulers, bureaucracies and parties is one of mutual dependency with each sharing the other's domain (Weber 1968). In this chapter, however, the intention is to listen in on the conversation between the tribunes and their government from the street side of the interaction.

Urban movements can take a variety of forms, and although we tend to speak more often today of 'urban social movements', because this term has been so closely associated with the 'new social movements' that emerged in the late 1960s, in order not to imply too limited a periodisation, we will mostly use the generic term in this chapter. In the pre-modern period the distinction between what

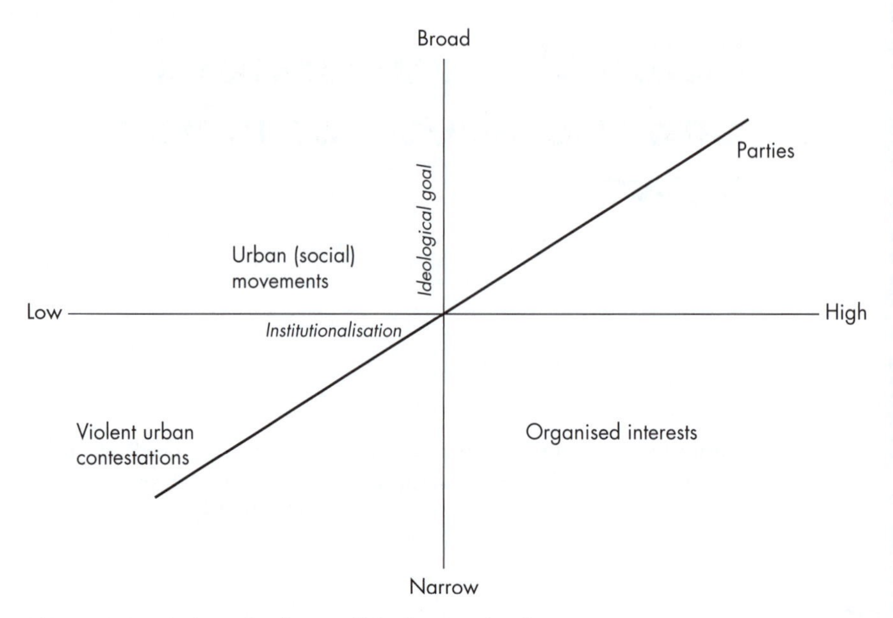

Figure 6. Varieties of urban political organisation.

constituted an urban movement as opposed to an 'organised interest' or 'party' is not easy to make, although even contemporary definitions often elide some or all of these terms. Nevertheless, and in light of our discussion in the previous chapter, it is useful to think of violent contestations, urban movements/urban social movements, organised interests and parties as existing along a continuum of ideological broadness and institutional thickness (see Figure 6).

Mass political parties are a quite recent and specialised form of urban political movement that mobilise and articulate the collective consciousness – the *conscience collectif* in Durkheim's phrase – while often acting at the same time as legislators and governors.

Such political organisations have not appeared *ex novo* but, depending on their ideological configuration, have evolved from elite factions surrounding a royal court or legislative body (curia, council, senate, tribune or parliament), or from collective organisations based around the defence or promotion of a particular social class, religious confession or territorial culture. That class struggle has often given form and purpose to urban movements is not an insight limited to Marxist historiography. In Livy, for example, we find that the ancient Roman polity is essentially characterised by the rivalry between rulers and ruled – the patricians and the plebs – even though we might have reason to question the claim that an organised plebeian movement suddenly emerged in 494 to challenge the authority of the Senate (Raaflaub 2005: 11).

Certainly the archaeological and historical record shows that there were violent social conflicts that afflicted the Etruscan cities of present day central Italy in the fourth and fifth centuries BCE, though we have to rely on indirect reports from later Roman sources in order to identify what lay behind these feuds. However, historians of antiquity and the Middle Ages are increasingly finding common ground in their analysis of 'the conflict of the orders', with parallels being drawn between the organisation of the commune in the medieval Italian city-states and the constitution of early Rome (Raaflaub 2005: 15). A significant difference, however, was the enormous growth of the slave economy under the Roman republic, which by 43 BCE had seen the slave population rise to some 3 million against a free population of 4.4 million (Anderson 1974: 62). From these captives of war and colonial expansion emerged the artisans (some 90 per cent were former slaves) who as *cittadini* were eventually to constitute an important element of the medieval economy after the fall of Rome.

In the thirteenth and fourteenth centuries the Italian city-states expressed their allegiance either to the Holy Roman Empire or to the Papacy through adherence to the Ghibelline party (which was loyal to the Hohenstaufen dynasty of Swabia in present south-west Germany and to Frederick Barbarossa's imperial challenge to the Pope) or to the Guelph party (the Italian form of the Welf ducal dynasty of Bavaria which remained loyal to the papacy). We can certainly identify some of the features of modern urban political movements in these great medieval cleavages between Papacy and Empire such as the need to forge alliances or defensive coalitions against a shared threat. The key aspect of these movements (which Dante's Inferno sees Mosca dei Lamberti suffer in hell for) is that Europe experienced a sustained mobilisation of resources (both human and material) over several centuries on the basis of schism or division. The meta conflict between a succession of Emperors and Popes found realisation at an urban level where often quite unrelated conflicts could be attached to the more global cause and given political expression.

The important work on the origins of the European party systems by Seymour Martin Lipset and Stein Rokkan explains the key role played by cities and towns during the latter half of the nineteenth century in producing a distinctively national bourgeois political identity that gave rise to very similar urban political movements from Norway to Naples (Lipset and Rokkan 1967). In other parts of the world that had been subject to colonialism political parties provided a vehicle of support and often of opposition to the colonial regime, which in the latter case gave rise to metropolitan-based nationalist parties. In Russia and China, throughout most of the nineteenth century the power of dynastic monarchy, rigid social hierarchies and the lack of an autonomous bourgeoisie inhibited the types of political self-organisation seen in Europe and its colonised territories (Moore

1966). In both cases an urbanised bourgeoisie succeeded in taking control of the state apparatus only fleetingly, before the arrival of communism and the monopolisation of the political life of Soviet and Chinese cities by the party-state.

Although cities have been essential for the formation and establishment of bourgeois hegemony they have also provided a space for 'the dangerous classes' – the poor and the dispossessed – at times to challenge authority and at others to do its bidding by persecuting aliens, outsiders, conspirators and traitors. The mob that cheered the procession of Louis XVI to the guillotine was the same one that celebrated the execution of Robespierre (Fife 2004). Yet, despite widespread support for national self-determination in the midst of an unpopular war, neither the civilian workers nor the enlisted servicemen of Dublin proved as receptive to the appeal of the insurgents during the Easter rising of 1916 as their Russian equivalents during the Bolshevik uprising in Petrograd a year later (Burdzhalov 2001; Townshend 2006).

As we observed in the preceding chapter, insurgent movements were often forced to abandon the city for the country where they could establish their own hegemony over the rural population and make use of the lack of communications infrastructure and the difficulty of the terrain to protect their forces from attack by the metropolitan regime. Cities have not therefore always been comfortable refuges for counter-systemic forces and movements but they have been instrumental in generating the ideologies, leaderships, support bases and material resources necessary to the formation of political oppositions.

Violent contestations against urban and national rulers, and among and between competing ethnic, religious and status groups, have often found their expression in the city. To the casual observer such eruptions of popular discontent can appear to be random, anarchic, spontaneous moments of collective madness that are symptomatic of the pathological nature of metropolitan life. In reality, as we saw in the case of Colombia, few sustained acts of urban collective violence or protest are without complex and historically and spatially structured causes. If it has a political origin or objective, collective political behaviour will eventually assume an organised form whether it be the loose association of a social movement, the more formalised and policy oriented vehicle of the interest group, or the state–society straddling 'modern prince' that is the political party (Gramsci 1971).

The development of urban politics in the United States, Canada and the United Kingdom

As Malcolm Low points out, urban political theory has had relatively little to say about party politics except as a sub-set of pluralist theory in which parties are considered alongside other collective actors as agents of urban political mobilisation. Yet since the origins of modern democracy political parties and their personnel have been essential to the provision of candidates and campaign staff in local elections, they supply elected and non-elected city officials from mayors to magistrates and school board members, they develop and implement policy programmes, and coordinate the activities of urban government with other political scales (regional, national and international) and with relevant actors in the local community – business associations, faith groups, trade unions and NGOs (Low 2007: 2653–4).

The study of political parties has been weakly associated with their spatial character (except at the national scale) and much of what passes for theory on the organisation, function and goals of political parties within cities derives from the rational-actor or neo-classical economics inspired work of writers such as Schumpeter and Downs, whose default model was that of liberal, competitive, capitalist democracies such as the United States, which then, as now, accounted for a minority of the world's urban population. As Tom Bottomore wrote in the introduction to *Capitalism, Socialism and Democracy*, the Schumpeterian view of democracy is one in which competition for political leadership exists 'as a form without any definite substance in the way of social or political ends', thus excluding a historical analysis of democratic politics that could help to show why it has flourished in certain locations and periods and stagnated or diminished in others (Bottomore in Schumpeter 1987: xviii).

The Cold War distinction that Schumpeter drew between the egalitarian despotism of centralist socialism and the more familiar world of the rational, dispassionate, electoral consumer and the power maximising purveyors of 'political goods' has led to a range of assumptions about the 'decline of party', the rise of 'networked governance' and technocracy that often fails to investigate in sufficient detail the so-called 'golden age' of party during the expansionist phase of Fordism from the 1920s to the 1960s.

However, as we know from the work of Robert Michels, the hierarchical, centralist and leader-oriented structures of modern political parties are almost always the result of the institutionalisation of more fluid and disparate political movements (Michels 1966). From the 1840s *Gangs of New York* depicted by the novelist Herbert Asbury (and later by Martin Scorsese in the movie of the same name), in which Nativist Protestants and Irish Catholic immigrants battled for control of New York's crime-ridden Bowery, came the machine-politics of Tammany Hall, which by the 1860s had become an Irish Catholic fiefdom and an important power base for the Democratic Party both within the state of New York and nationally. Elections were bought and sold in a manner that had little to do with democracy and everything to do with the ability of ward bosses to mobilise their supporters in favour of one or other faction. As Patrick Moynihan writes:

> Tammany was not simply a concentrated version of the familiar American municipal power structure in which an informal circle of more or less equally powerful men – the heads of the two richest banks, the three best law firms, four largest factories, and the chancellor of the local Methodist university – run things. Power was hierarchical in the machine, diffused in the way it is diffused in an army. Because the commanding general was powerful, it did not follow that the division generals were powerless – anything but. In just this way the Tammany district leaders were important men; and, right down to the block captain, all had rights.
>
> (Moynihan 1993)

If, as Moynihan claims, the roots of New York city politics can be traced to the power hierarchies and practices of the southern Irish village, it is curious that so much political science appears to regard electoral competition as the rational pursuit of individual self-interest, free from any such ethnographic or cultural particularities. In large American cities such as New York, Boston and Chicago, politics assumed a quasi tribal character as immigrant communities strove to achieve recognition and obtain resources from local elites who in turn needed clienteles, volunteers and donations to fill ballot boxes, fund electoral campaigns and staff their offices.

However, as the United States witnessed the emergence of a concentrated urban population in the latter part of the nineteenth century, its residents became both reliant on the products of industrialisation and essential to the servicing of the state and economy to which industrial capitalism had given rise. In the view of Carl Degler, while Tammany-style politics persisted in New York, Boston and San Francisco, the new urban demography began to favour political parties that were capable of aggregating interests on a national scale. This helps to explain why during the progressive era, the strongly unionist Republican Party appealed to the urban constituency because it was prepared to use federal resources and powers to impose protective tariffs, to provide land grants and loans to rail road companies and to pledge national funds to support public schools. Even Friedrich Engels, who was no Republican sympathiser, was forced to admit that 'through the protective tariff system and the steadily growing domestic market the workers must have been exposed to a prosperity no trace of which has been seen here in Europe for years now' (Degler 1968: 102–3).

In Los Angeles, progressive-era politics was characterised by a series of protracted fights between progressive populists and the private utility companies who fought municipalisation tooth and nail (see Chapter 9). The city's senior officials, through their oversight of the modernisation and development of the city's infrastructure, became a hybrid of politician and administrator to the extent that 'bureaucratic power becomes the dominant expression of an "officially based partisanship" as opposed to the "spoils based partisanship" of a party run polity' (Schiesl 1977: 111).[17]

The enfranchised city

Most American writing on urban politics in the twentieth century took the empowerment of cities for granted. City Hall may have been in the shadow of Tammany Hall and the political machines of the great American metropolises, but there was no real dissent from the idea that the key to urban power was to be found within the boundaries of the city limits (Judd and Swanstrom 2008: 53). This assumption was not too surprising given the dominance of American political scientists and historians in the field of urban studies. The notion of 'the self-governing community' was seen as both a constitutional entitlement and as perhaps the most valued aspect of the American polity, which undergirded its democratic values, ensured protection from an over-mighty federal administration and facilitated civic engagement (Short 2001: 271).

Certainly, throughout the nineteenth century and well into the twentieth the United States was uncommonly open (compared to most other nations) to the formation of new municipalities and corporations to such an extent that 'the right of cities to self-rule' became as important a principle as individual freedom (Judd and Swanstrom 2008: 5). Permissive state municipal incorporation laws required remarkably low population thresholds in order to acquire the right to tax, to educate and to police, with the consequence that developers, homophiles and those seeking shelter from the costs of big cities could easily create a city in their own image (Short 2001: 274).

However, the reality of the contemporary US metropolis is rather different. As Short points out, more than 50 per cent of Americans live in cities of over 1 million people, and yet these rapidly growing metropolitan expanses are characterised by political fragmentation where 'a balkanized arrangement of numerous municipal governments and various school districts' prevails (Short 2001: 274). In other words if one were looking for examples of uniform, comprehensive and democratically accountable government, America's great cities are probably the last place one would expect to find them.

Although Canada's urban political development followed a somewhat different trajectory to that of the United States, some of the structural weaknesses identified in the American context can also be found north of the border. Leibovitz argues that the evolution of the Canadian local state needs to be understood in terms of its British municipal origins and the influence of the United States on the country's subsequent constitutional development. In particular, the complex intergovernmental relations that have resulted from a decentralised federal system have been accompanied by a territorial fragmentation that has led to significant economic and political disparities. While rapid and recent urbanisation has created a mismatch between spatial development and local political institutions with increasing social and ethnic diversity posing a challenge to traditional modes of urban governance (Leibovitz 1999: 200).

The net result of these trends is that cities have found themselves in a relatively weak position vis-à-vis the provinces and the federal government with the consequence that despite being a highly urbanised population, 'urban issues in Canada rarely became national issues' (Ley 1991: 181); while, since the abolition of the Ministry of State for Urban Affairs in 1979, 'urban affairs have remained the almost exclusive responsibility of provincial governments' (Keating 1991: 149).

The disarticulation of municipal politics from provincial and federal government is further emphasised by the fact that in order to tackle a widespread corruption problem in city administrations at the turn of the twentieth century, municipal council elections were contested by independent candidates without formal affiliations to political parties with the exception of Montreal, and even here the municipal parties were not linked to provincial or federal parties (Boudreau 2005: 99).The depoliticisation of Canadian local politics did not result in greater electoral participation, however. In Vancouver, for example, from the mid-1880s to the mid-1990s voter turnout in municipal elections has rarely exceeded 50 per cent, a pattern that is consistent with many other Canadian cities over this period (Stewart 1997: 169).

In the United Kingdom, the history of autonomous local government can be traced back to the emergence of chartered merchant towns and cities – most notably the Corporation of London – in the early Middle Ages (De Krey 1985). The incorporation of Wales into what was to become Britain in the thirteenth century followed by Ireland in the seventeenth and Scotland in the early eighteenth century gave rise to a uniquely nationalised polity in which the monarch exercised a tight grip over the county towns through his sheriffs, the bishops whose appointment was and remains a royal prerogative, and the magistracy who were drawn from the ranks of the local gentry and thus constituted the bottom rung of the pyramid of aristocratic territorial power of which the crown formed the apex (Palliser 2006; Chapter 7). Local elites enjoyed a greater degree of legislative and political freedom in Ireland and Scotland even during the maximal period of national state consolidation in the late nineteenth and early twentieth centuries, but the unitary character of the British state was given strong territorial articulation through a series of local government reforms dating back to the 1830s which aimed at reforming the patchwork and threadbare pattern of urban government that had existed since the Elizabethan era (Young and Garside 1982).

Subsequent Victorian reforms of urban government such as the introduction of compulsory elementary education, new transport, sanitation and sewage authorities, and the consolidation of metropolitan government in major cities such as the creation of the London County Council in the 1880s, emerged from the enormous transformations in urban society that resulted from industrialisation and the growth of empire. The rapid growth in the local state with its legions of professional engineers, architects, superintendents, school inspectors, social workers and town clerks had its roots in the utilitarian reforms introduced by Bentham and Chadwick in the previous century (Jones 2000). The need to monitor, control, educate and tax an increasingly expanding urban population required new institutions of territorially based government that, in order to be incorporated into the nation-state had to be subject to statutory and executive control, while at the same time permitting a measure of defensive self-government that the financially pressed urban bourgeoisie secured through their control of the rate-payers' franchise (Young 1975).

Equally, the emergence of an organised labour movement and its political representative organisations gave rise to 'municipal socialist' local administrations in cities such as Glasgow and London which were to find their echo in many parts of Europe. The conquest of local state power by radical working class parties provided opportunities to engage in a measure of social redistribution, collective ownership and international solidarity, which exemplified at the urban level what might be possible at greater national and even international scales (Parker 2001a; Parker 2001b). In the following chapter we consider how the organisation and operation of London government has been subject to both welfarist and redistributionist agendas and in more recent years by pressure from neo-liberal forces at the level of the state and from regional and international business coalitions.

But before looking at the operation of government, in the following part we consider how organised interests have come to be associated particularly with pluralist-type urban politics.

Organised interests

What constitutes an organised interest is just as relevant and complicated a question for political science in general as it is for the study of urban politics. According to an influential study on the formation of interests in western Europe, in pluralist-type societies such as the United States and the United Kingdom organised interests can be described as 'the unmediated demands of socio-economic groups themselves' (Berger 1981: 5–6). This is not the case in many other parts of the world, however, where parties tend to dominate the political field, often recruiting sectoral interests to their own purposes. In continental Europe, for example, both Marxist and Christian Democrat parties have played a hegemonic role in harnessing labour organisations, cooperatives, youth associations, women's groups, and leisure and cultural organisations to the broader goals of the political movement with the result that civil society organisations tend to correspond to the major political alignments in a manner that the Dutch political scientist Arend Lijphart in a famous 1960s study referred to as 'pillarisation' – or *verzuiling* (Lijphart 2008). In this model religious or class-based sub-cultures provide parallel civil society organisations, which are often coordinated through a dominant political party.

In the United States where research on interest group politics in American cities dates back to the 1930s (Lynd and Lynd 1937), even in the pioneering post-war work of Hunter (Hunter 1953) and Dahl (Dahl 1961) there was no clear indication of how organised interests influenced local decision-making processes (Cooper, Nownes et al. 2005: 206). Subsequent studies tended to emphasise the one-sided nature of organised interests, and identified business elites as being capable of both determining the policy agenda, or even keeping certain policy options (such as pollution control) off the agenda entirely (Bachrach and Baratz 1962). Other authors, such as Paul Peterson, have suggested that city managers are so thoroughly conditioned by corporate requirements in terms of their taxation and spending priorities that to talk of business interests is to give the mistaken impression that there are other credible policy alternatives to a pro-business agenda (Peterson 1981). However, while few urban analysts deny that business interests predominate in the American city, it cannot be said that labour unions, community and neighbourhood organisations, faith groups, think tanks, city-oriented lobbies and other non-business interests have no influence on the urban policy process, or there would be no reason for many of them to exist (Parker 2004: 206).

The problem that confronts the would-be analyst of these types of political organisation is how to distinguish between the mobilisation of collective interest and collective mobilisation more generally. In the case of labour unions one can identify both a movementist function in terms of taking direct action such as strikes, go-slows, demonstrations and acts of solidarity such as boycotts; and a lobby function in the form of full-time officials seeking to persuade national and locally elected representatives and their officials to enact amend or revoke legislation or government policy in favour of their membership. Rather than attempting to categorise organised interests in terms of their exclusive political function, it is better, as Figure 6 aims to show, to think of interest organisations as intermediary collective political actors that lack the institutional complexity, range and broader political objectives of parties, but nevertheless have a more focused, professional and instrumental relationship to legitimate and established power than their activist driven social movement cousins.

The Urban League in Chicago

The city of Chicago, besides being the home of the United States' first black president, has a reputation 'as a hotbed of community organizations as well as "a union town"' (Sites 2007: 2642). Chicago is also what Nelson Algren described as 'a city on the make' and one of America's most important business centres – its long established stock exchange being second only to New York in terms of the volume of trades. Like New York, Chicago has a historical reputation for 'boss' or 'machine' politics centred around powerful political dynasties and networks of local power in which corruption, clientalism and crime have often played a significant role (Simpson 2001).

Chicago experienced a considerable in-migration of poor African American workers in the decades after the First World War who were fleeing from the economic hardships and discrimination of the south to fill the expanding demand for labour in the city's meat-packing, manufacturing and distribution industries. The National Urban League (NUL) was originally founded in 1910 when it was known as the Committee on Urban Conditions Among Negroes to assist the large numbers of African American city dwellers in finding jobs, accommodation and access to public resources at a time when Jim Crow and racist discrimination from both employers and labour unions were rife.

The NUL subsequently established its headquarters in New York City where it 'spearheads the non-partisan efforts of its local affiliates', of which there are over 100 located in 35 states and the District of Columbia that provide direct services, advocacy and research to more than 2 million people nationwide. The League organised boycotts against firms that refused to employ black workers, pressured urban schools to provide better vocational opportunities for young people, persuaded Washington officials to include black people in New Deal recovery programmes and strove to win membership for African Americans in previously segregated labour unions.

The tax-exempt status of the NUL meant that it was forbidden to engage in protest activities, but it was an important ally of the civil rights movements and at its New York headquarters the plans were drawn up for the 1963 March on

Washington. The anti-poverty legislation that was passed in the ensuing decade was strongly influenced by NUL executive director Whitney Young Jr's call for a domestic Marshall Plan to combat the great social and economic divide between the country's white and black populations.

The Chicago Urban League (CUL) was founded not long after the national organisation in 1916 'to help rural African Americans migrating from the South in unprecedented numbers adjust to urban living'. Its former president Edwin C. ('Bill') Berry described Chicago as 'the most important city in race relations in the world'.[18] The CUL has always emphasised the importance of 'economic empowerment' – 'the Chicago Urban League supports and advocates for economic, educational and social progress for African Americans through our agenda focused exclusively on economic empowerment as the key driver for social change' – which it promotes through initiatives such as the Entrepreneurship Center in association with Northwestern University, and programmes such as nextONE which are designed to accelerate the growth of high-potential African American-owned firms.[19] As the CEO of the CUL, Cheryl Jackson, put it in 2007,'The Chicago Urban League is getting out of the social services business and will focus exclusively on economic development. Moving forward, we will lead with an economic agenda to drive social change.'[20]

The CUL's transformation from a civil rights movement aimed at structural political change, into a major social enterprise which has established numerous community development corporations (CDCs) across the United States, which, according to DeFilippis, in order 'to be successful . . . must adopt an explicitly entrepreneurial set of goals and practices' (DeFilippis 1999: 982), serves to illustrate the argument that collective political actors and institutions cannot be separated from the socio-spatial context in which they operate.

Urban movements

Not the least of the problems that faces the analyst of urban social movements is to decide to what extent if any the collective political action under scrutiny can be described as urban – in terms of the content of the movement's goals, the nature and geographical constituency of its membership and leadership, the jurisdiction of the power holders to whom the movement's demands are targeted and so forth. Cities are not the only place where organised protests or direct action occur – poor farmers, landless labourers and indigenous populations faced with a loss of land and natural resources are often forced to engage in resistance that goes unrecorded and unnoticed far from the centres of regional and national political power. However, it is significant that even movements for regional autonomy such as that of the EZLN[21] in Chiapas, Mexico targeted Revolution Square in the capital city as the destiny of their national mobilisation in 2001, just as Martin Luther King's civil rights protestors converged on the Lincoln Memorial in Washington DC at the culmination of the civil rights movement in August 1963 as the symbolic site of national renewal and reconciliation (Giugni 1999: xiii).

The streets and squares of the city have long been a theatre for the oppressed as much as a *mise-en-scène* for the pomp and circumstance of both secular and religious elites. However, the city's function as a venue for the collective demonstration of claims against or manifestations of established power does not help us to distinguish between *sui generis* movements that happen to take place periodically in cities, and 'urban movements' that owe their origin and *raison d'être* to the city itself. In *The City and the Grassroots*, Manuel Castells characterises urban social movements as being involved with collective consumption demand-making, community cultures and self-management or 'autonomy' (Castells 1983 cited in Schuurman and Naerssen 1988: 2). While, in his later work, Castells seeks to emphasise the identity-based nature of many urban-based movements, moving away from strictly class-based or collective consumption-based cleavages to a more voluntaristic and pluralistic conception of collective political action (Castells 1997). We will explore the relationship between social identity, urban space and power in greater detail in Chapter 8, though it is worth noting here that in recent decades there has been a coming together of what one might call traditional urban social movements (such as labour unions and community organisations) alongside more 'post-materialist' movements such as climate change protestors, anti-capitalist/anti-globalisation groups, and pro-poor coalitions including NGOs. While not necessarily exclusively urban in their outlook, such movements have a large urban constituency in terms of the populations they claim to represent and very often a strong urban activist base that follows the political and business elites' global caravan of high level summits from global city to global city.

According to Pierre Hamel and colleagues, '[u]rban movements have usually been thought of as heavily vested in micro-, as opposed to macro-, social processes, underscoring the local characteristics of a specific urban political economy'. In other words, there has been a tendency to bracket urban movements with local, municipally oriented protests and community activism, which are closer to the character of organised interests than the broad, value-led, socially transformative collective movements more typically associated with 'new social movements' (Hamel, Mayer and Lustiger-Thaler 2000: 1). In a recent article Walter Nicholls argues that cities are important as a separate area of social movement research because they 'stimulate the formation of diverse groups with strong ties . . . enabling actors in these groups to pool and concentrate high-grade resources to address particular concerns' (Nicholls 2008: 841–2). Similarly,

William Sites considers how recent work in the field of urban movement research is revealing the ways in which 'urban mobilizations may be networked and scaled' (Sites 2007: 2633).

Although Hamel and colleagues are clearly right to stress the increasingly global dimension and activism of urban movements, it nevertheless remains the case that many urban movements are engaged in various ways in demanding the right to the city – whether in the form of affordable housing, the 'regularisation' of refugees and migrants, access to decently paid employment, public services such as education, health and welfare, or to a safe and pleasant environment. In focusing on the examples that follow, the aim is less to provide representative examples of urban movements since there is simply too much variation between different cities and movements to make this feasible. Instead the intention is to emphasise the importance of the economic, social, spatial and political context within which popular urban movements operate and to show how vertical relations of power can facilitate or constrain the emergence and development of urban movements.

Urban movements in the Philippines and Brazil

In the cities of the Global South, the mobilisation of the urban poor is often the result of even more desperate motives than those identified by Castells. With 1 billion of the world's urban population now existing in informal settlements or slums on the outskirts of the world's major cities (Davis 2006: 19), very often such communities face a daily battle for survival against developers, government authorities and criminal gangs (see Figure 7). In the South, 'agrarian transformation, industrialization and urbanization' have been the major instigators of this huge rise in the population of cities and the increased competition for jobs, housing and resources (Vellinga 1988: 154). However, the response to the pressures imposed on a historically exploited and under-developed urban South have varied considerably depending on the practical opportunities for mobilisation, the possibility for political transformation at the metropolitan, regional, national and international scale; and the long term translation of movement goals into concrete policy outcomes.

Manila

In the case of the Philippines, squatter communities in the Metro Manila region became highly organised in the 1970s and 1980s when the government attempted to evict the squatters and relocate them 45 km outside the city in order to build a

Figure 7. A sprawling informal settlement on the outskirts of Metro Manila. © Fuzzed/Dreamstime.com

new container port in the Tondo Foreshore area, which was the largest of Manila's slum quarters (Van Naerssen 1988). Founded by a group of Catholic and Protestant priests who had worked with the local people for many years, the Zone One Tondo Organization (ZOTO) 'borrowed much of its spirit and methods from the labour unions, using tactics available to the poor such as marches and rallies'. ZOTO also 'tried to educate people through workshops and reflection sessions but believed that the best learning came in and through action' (United Nations Economic and Social Council for Asia and the Pacific 2009). ZOTO subsequently developed into a Manila-wide 'territorially based federation of poor organizations', which managed to survive the imposition of martial law by the Marcos regime. Although ZOTO was unable to prevent up to 33,000 families being resettled outside Manila, the remainder of the 180,000 succeeded not only in holding onto their homes but, in negotiations with the sponsors of the port project (the German government and the World Bank), ZOTO won an undertaking that the remaining families would receive land titles and that their neighbourhoods would be upgraded (United Nations Economic and Social Council for Asia and the Pacific 2009).[22]

Faced with the government's failure to deal with the housing crisis of the urban poor, district-level grass-roots movements such as Ugnayan and Alyansa decided to merge with other municipally based groups across Metro Manila in order to campaign more effectively at national level for community-led planning, an end to the demolition of squatter housing together with broader economic aims such as lower prices, employment growth, state control of basic industries and rural land reform (Van Naerssen 1988: 208–9). Today ZOTO is a well established NGO whose programme 'Rebuilding Lives and Reshaping Better Futures', 'a comprehensive training program that builds the leadership capacities of young people', has attracted donor support from several overseas countries.[23]

The successful emergence of community-based squatter organisations as fully fledged NGOs should not lead us to overlook the fact that, important though self-help grass-roots movement are in combating and alleviating poverty, as Shatkin observes:

> in the context of globalization, pressures and incentives to attract corporate investment and upper income people has led in many cases not to more competitive and efficient government, but rather to the empowerment of elite economic actors at the expense of community groups. In the Philippines, numerous studies have demonstrated how local political leaders, motivated by vested interests in local land development and economic growth, systematically undermine efforts at independent community and labor organizing through intimidation, cooptation and violence.
>
> (Shatkin 2007)

Urban movements are learning to network horizontally on a regional and increasingly worldwide scale while, for some such as ZOTO, the establishment of political and economic relationships with domestic and foreign governments as well as transnational agencies such as the World Bank (however asymmetric) highlights the increasingly global political arena within which such community movements are now operating.

Porto Alegre

There can be few cities in the newly emerging economies of the Global South that represent the ideal of empowered local urban communities more comprehensively than Porto Alegre in Brazil. This city of some 4 million inhabitants has played host to the World Social Forum – a global alliance of NGOs, social movements, trade unions and political parties committed to alternatives to capitalism no less than three times. In 1988 Porto Alegre elected a left-wing Popular Front coalition government dominated by the Brazilian Workers Party (Partido dos Trabalhadores, or PT). Under its charismatic mayor, Tarso Genro, Porto Alegre embarked on an unprecedented experiment in participatory democracy, the central pillar of which was the administration's commitment to ensuring that its budget would not be determined by the mayor and his executive but directly with and through the people themselves. As one might expect from an avowedly left-wing political movement whose ex-leader, Luiz 'Lula' da Silva, is the former Brazilian president, the commitment to participatory budgeting was seen as an instrument for 'reversing the priorities' of public policy in favour of the poor (Gret and Sintomer 2005: 3). This was an urgent task for the government of a country that has some of the most glaring wealth disparities in the world (Thomas 2009: 73).[24]

The Porto Alegre city government made use of a recent federal reform which gave local authorities greater discretion in raising local taxes (especially property taxes), allowing the city's income to virtually double in the space of three years (Gret and Sintomer 2005: 54). Despite its revolutionary socialist origins, once in power the PT abandoned any aim of destroying 'the bourgeois state' and instead set about a radical renewal of public management. What was most significant about the participatory budget process from the point of view of urban movements and their influence on city politics was the city government's and the participatory budget council's (the COP) decision to give greater priority in terms of the division of the city budget to the level of popular mobilisation and the relative strength of the participatory movement within a given district.

One of the immediate effects of giving Porto Alegre's community organisations a genuine say in the control of their city was to change the number of districts from four to 16 in order to better represent the neighbourhoods with which local people actually identified.[25] This created an incentive for working-class participation in the poorer districts, and in particular promoted a greater degree of activism and self-management among working women. As a result, in the first few years of the participatory budget experiment the five highest priority districts received 65–70 per cent of distributed resources which produced 'fairly strong discontent in those areas least favoured by the mechanism' (Gret and Sintomer 2005: 56–7).

The Popular Sovereignty Network estimates that there are now some 600 cities around the world that have experience of participatory budgets, and Porto Alegre councillors have been invited to share their knowledge of grass-roots democracy with cities such as Toronto, where the income disparities between wealthy and poor residents are less stark but still evident (Boudreau et al., 2009: 133). However, even within Brazil and in Porto Alegre where this experiment in community-based collective democracy is at its most advanced, the role of social movements can often become marginal as legal–rational rights-based bureaucracies seek to transform advocacy groups into passive clients and label deviant and disruptive those movements and collective organisations that seek to challenge or undermine the status quo.[26]

Conclusion

Collective political action in cities takes a variety of forms and it is organisationally structured according to the nature of the goals and objectives that groups set themselves, by the opportunities and constraints provided by state actors and the resources and obstacles within civil society, and in particular by the nature of the political field in which parties, interest groups and movements have to operate within the urban/regional context (Parker 2006). This microclimate in which collective political associations operate is, however, constantly subject to weathering from macro- and meta-scale processes and trends. Thus if the dominant weather fronts in the nineteenth century were those of urbanisation and industrialisation, and in the twentieth century metropolitanisation, regionalisation and regulation, the long fin-de-siècle of the twentieth century has been associated with the integrating effects of globalisation in parallel with – at the level of the state – a seemingly contradictory neo-liberal drive towards decentralisation and deregulation.

Inevitably, the changing political field within which parties, interest groups and urban movements are obliged to operate has favoured certain types of organised politics and particular ideological objectives over others. As William Sites notes in the US context, in a period of neo-liberal state restructuring, the highly porous relationship that exists between instances of government and civil society can lead to state policies resulting 'from an established pattern in which state and party actors anticipate, accommodate, and therefore reinforce the short-term preferences of business and other economic actors' (Sites 2003).

This is no less true for urban labour movements which, faced with the increasing globalisation of capital, 'have to participate in local policies to attract "good jobs"'. As Greer argues, 'the globalization of capital and corporate organization leads to a localization or decentralization of collective labour participation' (Greer 2007: 193). Because cities tend to concentrate poorer populations and exhibit greater income inequalities and class variation than suburban and rural societies, they have also become sites of resistance against the commodification and marketisation of spaces and labour by state–capital alliances at every territorial scale (Leitner et al. 2007). It is interesting to note that in post-Fordist industrial societies capital has assumed an increasingly organised political form since the 1970s, with a significant growth in party political campaign donations by corporations and a proliferation in the membership and number of business lobbies in the 1980s and 1990s. By contrast, those who would have traditionally identified themselves on the political left are today more likely to be found in movements for social justice and environmental defence, in campaigns on behalf of refugees and migrants, the homeless, or in identity-based civil rights organisations focused around gender, sexuality, ethnicity and faith.

Whatever the form that political action in cities assumes, political institutions remain key interlocutors and agents of urban change. In the following chapter of this volume we move on to a more detailed examination of urban governance in terms of the organisation of city government and in Chapter 6 we consider the diverse scales and constellations of political and economic power that both constitute and orient the activities of the urban polity.

Further reading

Despite its predominantly North American focus, the Routledge *Urban Politics Reader* (Strom and Mollenkopf 2006) provides a number of essays of relevance and interest to the themes of this chapter, including the seminal 'How to Study Urban Political Power' by John H. Mollenkopf, Richard Croker on 'Tammany Hall and the Democracy', Paul E. Peterson on 'The Interests of the Limited City', Adolph Reed on 'Demobilization in the New Black Political Regime' and Michael Jones-Correa on 'Wanting In: Latin American immigrant women and the turn to electoral politics'.

Other key readings on urban movements include Manuel Castells, *The City and the Grassroots* (Castells 1983) and the chapter by Fainstein and Hirst on urban social movements in the first edition of *Theories of Urban Politics* (Judge et al. 1995). Although this latter volume is still a useful introduction to the topic, readers may find that the essays in the more recent edition of the volume edited by Davies and Imbroscio (Davies and Imbroscio 2009) offer a more contemporary and 'new politics' perspective on urban political movements. The volume by Rachel Abers (Abers 2000) on the Porto Alegre experience provides an in-depth account of Brazil's most well known experiment in radical democracy.

Ian Shatkin's *Collective Action and Urban Poverty Alleviation: Community organizations and the struggle for shelter in Manila* (Shatkin 2007) offers the most comprehensive account of the history and development of community-based organisations (CBOs) in Metro Manila since the Marcos dictatorship. For those looking for a sophisticated appraisal of the current field of urban political mobilisation, with a North American focus, the article by William Sites (Sites 2007) is recommended.

5 The government of cities

Creon: Is not the city held to be the ruler's?
Haemon: Thou wouldst make a good monarch of a desert.

Sophocles, *Antigone*

All I know is I am the goddamn mayor of the goddamn city of New York, the second most important elective office in the entire country, and you tell me I gotta lay here and suffer like every other schlub in the country?

Lee Wallace as The Mayor, *The Taking of Pelham 123*

Introduction

As we saw in the preceding chapter, organised interests and social movements had long established themselves as important protagonists even in the earliest examples of urban civilisations. It is only with the emergence of *representative* urban government and municipal elections in the early nineteenth century, predominantly in northern Europe and North America, that mass political parties emerged as important political actors in their own right – although some Latin American countries such as Venezuela enjoyed municipal autonomy earlier than many European states (Mörner 1993: 33). Local political elites and coalitions of interest built around religion, ethnicity, class and territory remained central to the operation of municipal politics, which meant that often party organisations were nothing more than vehicles for the election of rival political elites and for the recruitment of clienteles and the distribution of patronage.

In Chapter 2 we noted how, by the time that industrial capitalism had reached its full development in the Global North, cities had become significant centres of economic and political life in their own right, as well as regional and in some cases national and international command and control centres for a variety of civil and military governments and business and labour interests. These civil society manifestations of the emerging urban polity were nevertheless dwarfed by the rise in city-based government employment on both sides of the Atlantic in the early decades of the twentieth century. It was at this time that the first generation of social and political scientists began to develop an interest in the relationship between the state, economy and society – often choosing to adopt the city as a convenient and experimentally robust territorial scale from which to draw more general conclusions on the operation of political power within society as a whole.

Urban leadership and the management of cities

As we observed in the opening chapters of this volume, many different types of actors have been involved in the organisation and administration of cities and urban space, whether formally or informally. In Italy during the Middle Ages, a number of cities were ruled by a chief magistrate or *podestà* whose title derives from the Latin term for power (*potestat* or *potestas*). This office was revived by Mussolini during the period of the Fascist dictatorship when previously elected local mayors were ousted in favour of loyal party henchmen who took their orders directly from Rome (Angeli 2001). Similar urban rulers can be identified in the figure of the German *Bürgermeister*, the French *prefet* (Bäck et al. 2006) and in the captain-generals and governors of the European conquest in the Caribbean and Latin America, all of whom operated with the considerable economic, military and political powers conferred on them by the sovereign authority of the crown or the republic (Morse 1962).

As Daniel Kübler and Pascal Michel write:

> During the construction of the nation states in the 18th and 19th century, European cities were squeezed into the corset of national intergovernmental frameworks. No matter how glorious their past forgone, European cities henceforth occupy a subordinate position within national state polities. Their autonomy is limited by upper levels of government, such as regions, federate states, and the central state.
>
> (Kübler and Michel 2006: 221)

In the previous chapter we considered how the emergence of a competitive party system in the United States was not always associated with greater levels of democratic decision-making within the corridors of City Hall due to the growth and persistence of political machines that centred on powerful local figures and their clienteles. Where the national and local state bureaucracies were relatively underdeveloped and there was little tradition of collective decision-making at the municipal level such as in much of southern Europe prior to the 1980s, Olivier

Borraz and Peter John identify 'practices of strong individual leadership . . . with power and legitimacy vested in the office and person of the mayor who often acted as a broker between central and local actors and was often a powerful national politician as well as being the local boss'. By contrast, in northern Europe where the institutional and political landscape was less fractured, 'collegial processes appeared to predominate' (Borraz and John 2004: 110). In a similar vein, Goldsmith and Larsen maintain that the Nordic countries, because of the strong degree of ethnic, linguistic and religious homogeneity and the relatively small size of the urban settlements, operate in a consensual and corporatist mode of governance (Goldsmith and Larsen 2004).

This stress on the nature and capacity of urban governance, rather than the form of government, is exemplified by Prud'homme's claim that 'good management is a matter of people, but also of institutions'. In other words, in an age of globalisation and increasingly porous governmental scales, successful political leaders need to be more than the patrician clan chiefs that typified 'boss politics' in the nineteenth and early twentieth centuries. Instead they have to be able to coordinate and broker constellations of parties, interest groups, urban movements and the different scales and networks of government that coalesce within the metropolis (cited in Goldsmith 2001: 327).

This idea of urban governance tends to stress the agency of political actors rather than the economic and political structure within which different groups seek to raise and satisfy their demands as key to answering 'who gets what, when and how' type questions and has its antecedents in a series of landmark studies undertaken by political scientists during what we might call the adolescence of the American city.

Community power

Robert Dahl's now classic work on the history of urban government in New Haven, Connecticut, which has become the defining text for classical pluralist theory (Dahl 1961), was written against the prevailing elite politics orthodoxy that emerged during and after the savage years of the Great Depression. These pioneering authors were far more cynical about the operations of urban power than many of their post-war successors. Robert and Helen Lynd's pioneering study of 'Middletown' (in reality the Midwest town of Muncie, Indiana) concluded that Middletown's affairs were not decided by its citizens but rather by a tightly knit oligarchy whose control over the local community appeared unthreatened by the occasional advent of an election (Lynd and Lynd 1929; 1937). Floyd Hunter, who studied the city of Atlanta in the 1940s and 1950s, came to a similar assessment of elite rule in Georgia where economic wealth and political authority decided what policies and decisions should be taken on behalf of the largely voiceless and powerless majority (Hunter 1953; 1959).

Dahl set out to test the limitations of the elitist view by examining official and unofficial sources of office holding and key leadership roles in New Haven since the origins of elected government in the city, and through survey questionnaires and in-depth interviews. Dahl did find that a self-reproducing elite ran New Haven in the eighteenth and nineteenth centuries, but as the city's population grew in size and its demography and economy became more diverse he found that it was impossible to identify a specific locus of power in the city. The reason for this was not that inequality and unequal access to power and resources did not exist at the level of the individual, but rather that at the aggregate level the ability of different groups to mobilise themselves and to compete against or form coalitions with the incumbent regime ensured that no one section of society could hold permanent sway (Dahl 1961).

In contrast to the ecological view of group behaviour and organisation that had dominated urban power debates in (and to an extent outside) North America since the 1920s, the community power school that emerged in the 1950s incorporated both elite writers such as Hunter and pluralist decision analysts including Dahl and Edward Banfield, whose *Political Influence* (Banfield 1961), purported to show post-war Chicago in a different light to conventional 'machine politics' accounts of one-party cities (see Chapter 4). While elitists and pluralists may have disagreed over whether power was dispersed or concentrated in urban communities, there was a fundamental consensus that 'the rules of the game' were fundamentally the outcome of a free and open society, even if, as writers such as Bachrach and Baratz (Bachrach and Baratz 1962) showed, the political agenda was open to manipulation and undue influence.

'Community power' has been enormously influential in giving shape to the three most important currents in the study of contemporary western urban politics: pluralism, elite theory and urban political economy. This latter category includes a broad range of theoretical perspectives ranging from Marxism and neo-Marxism through to public/rational choice models. But, as Alan Harding writes,

> community power theorists were generally more concerned with power than they were with 'communities' or their relationship to 'places'. For Hunter and Dahl, the 'communities' of Atlanta and New Haven were interesting not so much in themselves but in so far as they could be treated as microcosms of the larger (US) society of which they formed a part. . . . The 'urban' focus of community power studies . . . was rather vague and based largely on convenience for the researcher and the assumption that empirical research findings derived at the level of the city could be generalised.
>
> (Harding 2009: 31–2)

As urban political sociology and urban political economy grew into distinct sub-disciplines in the 1970s, the focus of more critical scholarship on urban power began to shift away from a strict focus on the holders of formal electoral offices to the wider economic, social and political forces that had an interest in shaping the future development of cities.

The urban growth machine

Harvey Molotch first published his account of the city as a growth machine back in 1976 and, despite refinements, its basic premise – that the city is an urban investment vehicle whose value depends on the collective commitment of local and regional political elites, developers, real estate owners, local businesses and other powerful stakeholders to achieve economic and demographic growth – remains salient. In stressing the importance of understanding the entire canvas of the city as an investment space, Molotch and his later collaborator John Logan saw entrepreneurs as having a straightforwardly instrumental view of the corridors of power, access to which they use 'to generate growth for their metropolis as a whole and for their sections of it in particular'. In short, 'the city *becomes* a growth machine, and its custodians are the people who grease its wheels, refurbish its parts, and tweak its direction as the need arises' (Logan and Molotch 2007: xi–x).

Harvey Molotch's claim in his original article (Molotch 1976) that 'growth policy is not just one of many facets of local politics but, rather, the guiding concern around which governments are constructed' (Logan et al. 1999: 73), may have been a bold one to make at a time when state Fordism in many parts of the world was still in the ascendant. Today, however, rather than appearing far-fetched, this statement appears to have acquired the status of an orthodoxy. Over the past three decades the concept of growth has expanded far beyond the real estate market to include every visible and invisible measure of a city's economic potential – from the number of tourist visitors a city welcomes each year to the percentage of 18-year-olds graduating from high school to the amount of dot.com start-ups.

Entrepreneurial cities by their very nature are pro-growth, but what Molotch and later collaborators were able to show was that 'growth' does not happen spontaneously, and neither is relatively faster growth in one sector of the economy usually an accident of market conditions. In other words, 'urban growth has to be understood not as a function of urban necessity but as the target of political action' (Logan et al. 1999: 74).

This is why in order to understand the purpose and causes of urban economic growth we need to be able to understand the political apparatuses, processes and alliances that make particular types of growth possible. In the original formulation of Logan and Molotch growth machines arise when rentiers (developers, realtors, banks, investors) and associated interests (the media, universities, utilities, sports franchises, chambers of commerce) combine with urban decision makers in order to direct fiscal and planning policies towards growth. The resulting higher levels of economic activity generate more demand for urban land (for housing, offices, amenities, roads, recreation) and increased exchange value for the landowners and their clients. However, the authors recognised that not everyone was in favour of unrestricted growth – in particular those urban residents who benefit from the use values associated with non-development (more open space, lower levels of traffic congestion and pollution, less crime, more affordable rents and real estate prices, etc.) are likely to be in conflict with the exchange-value maximising objectives of growth machines (Jonas and Wilson 1999: 5–6).

According to one author, cities such as Chicago and Pittsburgh 'have been governed by "pro-growth" regimes since the middle of the twentieth century'. Both cities were older industrial cities that had experienced the loss of population and jobs in the period after the Second World War and they continued to be run by well oiled political machines that enjoyed 'significant clout in the city, region, and state' that sought to institute economic development policies 'geared toward the new service-based economy' (Ferman 1996: 19).

In the case of Pittsburgh the pro-growth regime comprising the city's economic and political elites led by the Republican financier Richard King Mellon and the Democratic Mayor David Lawrence was a response to the economic crisis, which had particularly affected the steel industry (Ferman 1996: 44–5). In Chicago the 'Daley machine' began to break down after the death of Richard J. Daley in 1976 as internal feuds, conflict with the national Democratic Party organisation and the machine's faltering control over the African American vote undermined the unity of the governing coalition (Ferman 1996: 33–4). As these examples show incumbent growth coalitions can be defeated when the accumulation of patronage and the self-serving distribution of resources and policy decisions leads to a crisis of legitimacy and the potential for electoral revolts. Replacement urban elites generally find an accommodation with *ancien régime* coalitions or try to construct their own – very few, as the fate of the progressive Kucinich administration in Cleveland, Ohio in the 1970s demonstrated, managed to face down the opposition of financial and corporate capital to business limiting policies for long (Clavel 1986: Glasberg 1988).

Anti-growth interventions

Excessive levels of urban–regional growth can often produce huge territorial and socio-economic disparities at the level of the national or regional state, prompting many national governments to introduce metropolitan containment policies both in terms of the physical territorial area that the city–region is allowed to occupy and in terms of the public sector infrastructure (schools, roads, hospitals, police) that the state is publicly prepared to subsidise. However, as the Organisation for Economic Co-operation and Development (OECD) notes, '[e]xperiences of containment policies in OECD countries (such as the one conducted in Paris in the 1960s, in Tokyo from 1959–2002, in London from 1965–79 and still currently implemented in Seoul since the 1970s) have provided mixed outcomes. There is little reliable data showing whether constraints on the growth of the major region actually displaced economic activities to other domestic regions'.

The OECD clearly regards such top-down attempts to interfere in the locational decisions of populations and businesses as a barrier to competitiveness and ultimately counter-productive since footloose capital that is frustrated in seeking an optimal location in the most dynamic regional economy in country A will almost certainly find a welcome home in country B or C (Organisation for Economic Co-operation and Development 2006: 97).

There also exist, as Schneider and Teske (Schneider and Teske 1993) point out, 'antigrowth entrepreneurs' from within the business community who have a vested interest in maintaining the status quo, either because they fear the loss of dominant market position or because they do not wish to see their particular 'in' to the decision-making machinery threatened by new economic entrants.

The proliferation of 'bottom-up' or 'grass-roots' campaigns to limit growth in recent decades has given rise to an increasing amount of research into their composition, funding, political strategy and effectiveness (DiGaetano and Klemanski 1999). Much of this research has been concentrated on American towns and cities where growth limitation policies are understood to emanate from local citizen action, which often happens in concert with local business interests (particularly in the area of limiting population growth) and is by no means exclusive to white-collar or high-status residents (Baldassare and Protash 1982). The pattern that appeared to be emerging in the USA in the 1970s and 1980s was one of a widening hierarchy identified by John Logan as 'a human political ecology' in which 'persons and organizations constantly seek to affect the growth process in order to maintain or create inequalities among places to their own advantage' (Logan 1978: 406).

Urban regimes

The growth machine model argues that while politics matters, it matters only insofar as the coalition goal of growth is sustained. As we have seen, however, not all coalitions between local interests and the local state are 'pro-growth', and politicians can and do choose their allies in order to promote a variety of policy outcomes. In an attempt to breathe new life back into the 'community power' model a group of predominantly American scholars believed they had identified an emerging form of urban governance which they referred to as an 'urban regime' (Elkin 1987; Stone and Sanders 1987; Mossberger and Stoker 2001). Urban regime theory (as this school of urban politics came to be known)

> represents an attempt at responding to some of [the] criticisms of the growth machine model by focusing more on the nature of informal coalition behaviour (regimes) and acknowledging the lead role played by city administrators in establishing such urban partnerships.
>
> (Parker 2004: 127)

Paraphrasing considerably from Clarence Stone's helpful summary of the key criteria of an urban regime, we would expect to find an 'identifying agenda' or a rallying call that pulls together different urban stakeholders often associated with slogans such as Atlanta's 'the city too busy to hate' or Chicago's 'the city that works'. The arrangements that exist between the city government and urban stakeholders 'are relatively stable – though not static and 'have a cross-sector foundation' that is embodied in a governing coalition. The 'identifying agenda' is not that of a single elite group but that of the coalition as a whole. Arrangements also tend to be informal insofar as the actors are not regulated by the formal procedures of the public authority and collaboration is voluntary and cooperative in nature. Finally, arrangements have to add capacity and attract resources that would not spontaneously have happened without the coalition and the identifying agenda being in place (Stone 2001: 21).

In emphasising the centrality of urban political elites as agents of coalition formation and network building, urban regime scholars sought to challenge the idea that city leaderships were passive facilitators of urban growth. Nevertheless, the distinction between 'urban growth machines' and 'urban growth regimes' can be a fine one, especially in non-American urban contexts where the local and regional state is almost invariably the lead partner (Harding 1994). However, even in the case of well-studied metropolitan polities such as New York City, researchers claim to have identified quite contrasting patterns of urban governance (see, for example, Mollenkopf 1992, 1995; Dreier et al. 2001; Sites and Judd 2002).

The new urban politics

The idea that urban political power rests exclusively in the hands of a strong political executive at the level of the local and regional state has come under increasing challenge from proponents of the scalar turn in urban research (as we shall see in the following chapter), while in the fields of sociology and geography there has been a more long-standing interest in the role played by managers, public sector professionals (planners, architects, engineers, economists, lawyers) and 'street level bureaucrats' (teachers, social workers, police officers) in the strategic organisation and distribution of resources 'on the ground'.

An early proponent of what has come to be known as 'the new urban politics' (Cox 1993; Hall and Hubbard 1996; Jonas and Wilson 1999) is the British sociologist, Ray Pahl, who argued that the non-random distribution of urban resources reflected the fact that there were important managers or controllers of local goods and services who were able to act as gatekeepers while still operating under the general imperatives of the market and higher state forms. Nevertheless, senior urban professionals have to work within the constraints set by those at the top of the urban hierarchy. In 'Whose City?' Pahl remarked that, '[t]he will of the community is mediated through the political process, so that those with the most power set the goals, which makes the planner simply the tool of the elite' (Pahl 1970: 206).

Further down the urban food chain, more recent work has focused on what Michael Lipsky has termed 'the street level bureaucrat' (Lipsky 1980). Lipsky's research brought into focus the activities of the urban service providers who actually make things happen to urban residents, workers and visitors or to the urban environment. They include visible public servants such as teachers, fire fighters, police officers, social workers, housing managers and so on. Lipsky considers such public functionaries 'policy makers' insofar as they exercise wide discretion over the lives of citizens and enjoy relative autonomy within their organisational hierarchy.

Typically, street level bureaucrats are faced with having to manage chronically limited resources, increasing client expectations and a lack of consistency, focus and direction on the part of higher-level bureaucrats who are meant to provide an institutional policy agenda. As a consequence, they are often required to 'muddle through' (Lindblom 1959) the policy and regulatory gap created by inadequate resources and the problem of too many rules, targets and policy promises by initiating self-generated policy work-arounds (Molotch 2009). In other words, discretion is the hand-servant of political authority wherever formal political power has to be exercised at a distance and indirectly.

Whether bureaucrats operate at 'street level' or at a higher level within the institutional structure of the urban polity, there emerged a criticism of the urban managerialist approach from more avowedly Marxist scholars such as David Harvey, and in particular Manuel Castells who believed neo-Weberian urban scholarship was ignoring the deep structures of embedded class power by attempting to claim that the actions of public servants could constitute a separate sphere of political power in its own right.

In his earlier work, Castells argued that 'relations between power and the city must be studied from the viewpoint of the ability each class develops in orienteering social organisations according to its interests, among which are its interventions in the sphere of the social practices of the city'. He later adds:

> We have been able to outline the way this sphere generally corresponds in *advanced capitalism* to the production, distribution and management of the means of collective consumption . . . Urban politics are . . . at the basis of the urban structure, but it is from the action of the state which the whole of society expresses its orientations and its relations of force.
>
> (Castells 1978: 174)

Although in his later work Castells resiles from the notion that urban power can only be understood as class power as mediated through the agency of the capitalist state (Castells 1985), the argument in *City, Class and Power* highlights a continuing tension between social theorists who regard the urban system as explicable only in terms of the operation of capitalism as a mode of production, distribution and exchange and those who see merit in empirically investigating the relative decision-making capacities of a variety of urban actors.

In the remainder of this chapter we aim to test some of these approaches and insights by focusing on the development and functioning of urban political institutions and the networks of governance in three markedly different metropolitan contexts – Greater London and the rapidly developing urban regions of China and Mexico.

Governing London

Until the advent of the London County Council (see Figure 8) in 1889 historians might question as Metternich did in relation to Italy whether London could be considered as anything more than 'a geographical expression', having been comprised until then of a number of parish authorities, vestry bodies, ad hoc commissions and boards – with only the Metropolitan Board of Works established in 1855 providing some measure of a city-wide infrastructure for sewage and the main thoroughfares (Travers 2004: 22–4). Today the capital remains a fractured

Figure 8. The former headquarters of the London County Council and Greater London Council. © Anthony Baggett/Dreamstime.com

but increasingly complex polity (see Figure 9) comprising 32 separate borough authorities, a metropolitan-wide Greater London Assembly and London Mayor's Office under whose auspices operate a separate transport authority (TfL), London's fire and civil defence, the economic development agency for London (the LDA) and numerous specialist bodies that deal with parks, culture, sport, voluntary associations and so on. The metropolitan police service (known colloquially as 'the Met') is accountable to both the Home Secretary *and* the Mayor of London in his or her capacity as Chair of the Metropolitan Police Authority. At the centre of this global metropolis is the City of London, 'the square mile' which has jealously guarded its political and financial independence for almost a thousand years. The Corporation of London, as it should be properly referred to, has its own police force and a unique electoral franchise dating back to the Middle Ages, which ensures that aldermen and the Lord Mayor are drawn from the ranks of the city's economic elite.

Whitehall, the district of Westminster where Her Majesty's Government accommodates its officials, also has a direct interest in the affairs of London. The various ministries of state from Culture and Heritage to the Home Office and Work and Pensions all have an input into how London is run – preferring increasingly to use non-departmental public bodies[27] (NDPBs) and partnerships

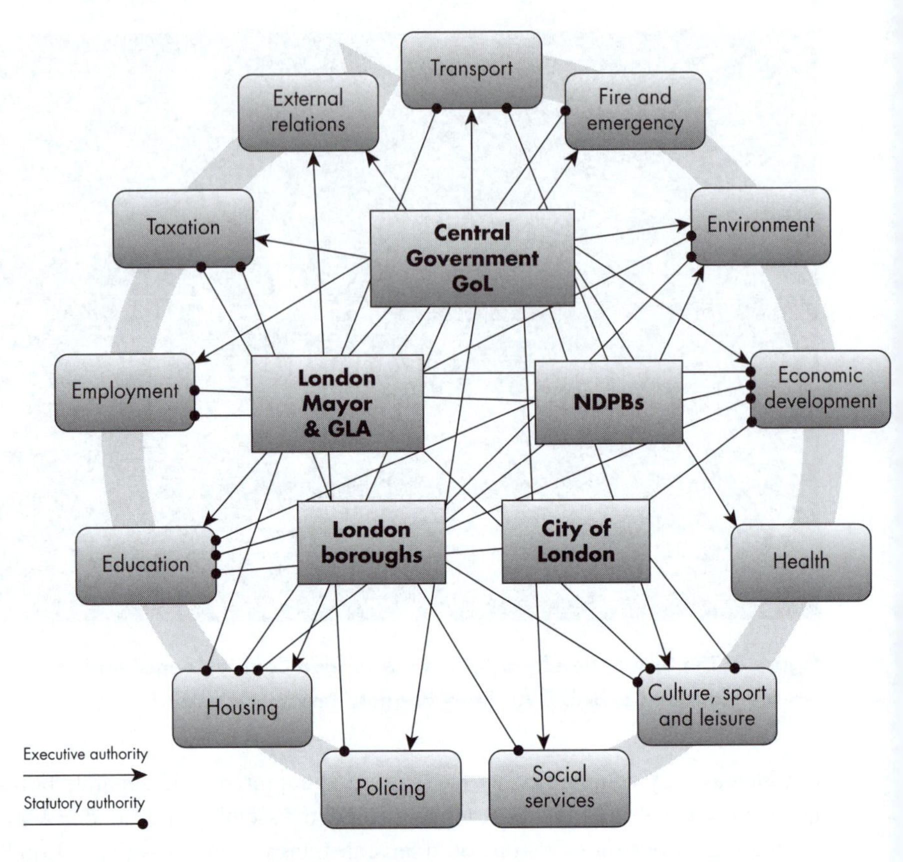

Figure 9. The web of London government.

of private, public and voluntary sector organisations to pursue their broader policy objectives. Many of the activities of the London regional authorities, NDPBs and partnership organisations are managed through the Government Office for London (GoL), which is one of nine regional offices in England. The GoL's major responsibilities include overseeing and allocating funding for the New Deal for Communities programme established by the Labour government, along with some 20 Local Strategic Partnerships, which are mostly targeted at disadvantaged London boroughs. The GoL has been involved in producing Local Area Agreements to ensure that national policy strategies 'are embedded locally', and portrays itself as being a key instigator of what the former British Labour government referred to as 'joined up government' – or facilitating 'single conversations across the government agenda' (Government Office for London 2006: 5). However, the Conservative–Liberal Democrat coalition government elected in May 2010 has since pledged to abolish the Government Office for

London as part of a new 'localism' agenda which is likely to result in the winding-up of the remaining English regional offices.

The Greater London Authority Act, which was passed in 1999, allowed for the introduction of a directly elected assembly and mayor for London following a 72 per cent yes vote in the London wide referendum in the preceding year. In a previous public consultation document (known as a 'white paper'), the national Labour government under Tony Blair called for 'A Mayor and Assembly for London' to re-institute a metropolitan government for London. However, this new authority was to be on the American model of a strong mayoral executive figure with considerable financial control and a relatively weak elected assembly whose powers were mostly limited to a scrutiny and annual budgetary approval function. The key London-wide authorities that come under the auspices of the Greater London Authority (GLA) are the Metropolitan Police Authority (MPA), the London Fire and Emergency Planning Authority (LFEPA), the London Development Agency (LDA) and Transport for London (TfL) which operates London's public transport system, road network and traffic management including the central London congestion charge.

The GLA and the Office of the London Mayor provide a useful starting point for those wishing to explore the capital's 'power map'. When Margaret Thatcher abolished the Greater London Council and several other major metropolitan authorities in 1986, she had already prepared the ground in the early 1980s for the de-democratisation of urban governance by creating a number of 'arm's length' publicly funded but corporately managed bodies such as Enterprise Zones, Training and Enterprise Councils and Development Agencies (Goldsmith 2001: 330). These pro-business quasi-public agencies were accountable to their ministerial sponsors, or to locally established boards that sometimes included representatives of local authorities, but were deliberately not meant to be accountable to the local communities in which they were situated.

Established in 1981, the most important and controversial of these new urban governance agencies was the London Docklands Development Corporation, which was granted exclusive authority to develop a new bespoke location for global capital on the site of the former London docks following the massive expansion in international banking and financial services that accompanied the deregulation of the City of London in the mid-1980s (see Figure 10). The history of the London Docklands Development Corporation is indicative of the type of institutional 'roll-out' neo-liberalism identified by Peck and Tickell (Peck and Tickell 2002). Initially it was greeted with hostility by the mostly Labour dominated local authorities that surrounded it on the north and south banks of the River Thames. But over time, as Tony Travers notes,

> the relationship steadily improved from being adversarial in the 1980s to a partnership approach in the 1990s as local Labour politicians changed their anti-development stance to 'a gung-ho pro-development' approach largely because they realised that strategic partnerships between the business sector, voluntary and community groups and other local authorities and NDPBs were the only way to access the considerable amounts of funding that were needed to regenerate local neighbourhoods.
>
> (Travers 2004: 40–1)

Having been at one time a notorious opponent of the City and big-business interests, Ken Livingstone's rapprochement with Tony Blair and the Labour government before and after his successful second election campaign was echoed by the mayor's enthusiasm for extending a London Docklands-type development further east along the Thames Gateway as 'the only way of sustaining London's continued growth and leading position in the global urban hierarchy' (Butler 2007: 760).

The increasing integration and dependency of the wider metropolitan region (known as the London Functional Urban Region or FUR) has not, however, been matched by an equivalent recalibration (to use Brenner's term) of sub-national governance (see Chapter 6). In the case of London, the FUR now extends across the 100 mile (160 km) radius envisaged by H.G. Wells from the Solent to the Wash (Sudjic 1995), a territory which had a census population of 12.5 million in 1991 – 10 times larger than the equivalent Parisian FUR and five to six times larger than the next biggest FURs of Greater Manchester and Glasgow–Edinburgh (Buck et al. 2002: 19). After a period of relative decline, in the late 1980s the

Figure 10. The London Docklands development. © Godrick/Dreamstime.com

population of the core city defined as the 32 London Boroughs began to grow again, amounting to some 3 million in the inner city boroughs and 4.56 million in outer London. In 2007 the capital city accounted for 12 per cent of the United Kingdom population, and provided 38 per cent of its natural population growth. The net loss of population to the wider functional urban region is compensated by new arrivals from overseas, many of them young migrants of working age, prompting the Greater London Authority to refer to London as 'the demographic engine of the UK'.[28]

As Buck and colleagues observe, the pro-growth business-led coalition London First, which represented many of the global companies with headquarters in the capital strongly endorsed the creation of a Greater London Authority 'primarily to advance the competitiveness agenda' (Buck et al. 2002: 319). The territorial 'new constitutionalism' of Labour in this period, which extended to the creation of separate parliaments and assemblies for Scotland and Wales, combined state restructuring with a limited restoration of local democratic control within the context of a growth coalition agenda in which non-market-based government alternatives were specifically excluded. In the new ruthlessly competitive world of transnational capital and labour flows, even political leaders such as Livingstone who had berated the previous Thatcher governments for being the servants of chauvinist capitalism, agreed that the 'competitive city' was a reality that had to be accepted and managed in the interests of an urban population that remained largely excluded from its benefits.[29]

Governing the urban system in China and Mexico

One of the enduring questions in the study of cities under socialist or communist state control is to what extent can we speak of a distinctly urban form of socialist administration as opposed to a general system of party-based rule that also included towns and cities. According to Szelényi it is possible to identify distinctly socialist patterns in the socialist countries of Eastern Europe in the period from the late 1940s to 1990 but this was less to do with socialist planning so much as 'the consequences of the abolition of private property, of the monopoly of state ownership of the means of production, and of the redistributive, centrally planned character of the economic system' (Szelényi 1996: 287 in Logan 2002: 6). With some notable exceptions, most 'actually existing' socialist societies had not experienced a prolonged period of democratic government.

China

As John Friedmann writes, 'Chinese cities were never corporate entities with their own legislative bodies, and never became cradles of democracy' (Friedmann 2005: 95). The imperial *yamen* system – a county-level government responsible for civic and judicial administration as well as the levying of taxes – was corrupt and widely despised, and unsurprisingly it was one of the first institutions to be disbanded after the creation of the republic. Urban government became somewhat more formalised under the nationalist Kuomintang regime, with the police force performing a super-regulatory role in major cities such as Beijing, but otherwise chambers of commerce, professional bodies, students and other associations were encouraged to provide their own ad hoc administration (ibid. 98–101).

Mao Tse-tung's idea of urban government was hardly more developed than that of his predecessor. Under Mao's leadership the Chinese Communist Party (CCP) began to organise the country's towns and cities into a series of work units or *danwei*. In a striking emulation of the factory towns found in different regions of continental Europe in the nineteenth century, the Chinese *danwei* was a walled compound organised around a state-owned enterprise or other institution, with its workforce housed in small apartments provided by the state at nominal rents. From cradle to grave workers would have no reason to leave the environs of the *danwei* as their needs, care and livelihood were all guaranteed. At the same time the 'anarchy' of the republic city was swept away by the Communists and replaced with an egalitarian feudalism, which by 1960 dispensed with the need for physical planning.

It is an open question as to whether, following the reforms introduced by Deng Xiaoping and his successors, the People's Republic of China (PRC) can truly be considered a 'post socialist' society given the continuing grip on power exercised by the CCP, and the PRC's continued commitment to the principles of Marxism–Leninism and the teachings of Mao Tse-tung. However, following the death of Mao in 1976, China's political leaders reversed two of the fundamental policies of the great leap forward and the cultural revolution – the pro-ruralisation strategy that had been a disaster for China's economic growth and the ban on private business activity which had also limited the country's productive potential. According to one study there were only 80,000 licensed private businesses just prior to the introduction of market reforms in 1978; 10 years later there were 30 million private businesses of various sizes and types, which represented the fastest growing sector of the Chinese economy (Wank 1999: 7). China's urban population rose from 11.2 per cent in 1950 to 27.6 per cent in 1990 to 39.2 per cent in 2000 and was predicted to reach 58.3 per cent by 2010 (Chan 1994: 153, 158).

The consequences for the government of China's towns and cities have been considerable and have led to what Wu and others have described as 'the reconsolidation of local power', with reforms such as the 1989 City Planning Act giving municipalities 'the right to prepare urban plans, to issue land use and building permits, and to enforce development control' (Wu and Yeh 2007: 119). However, as the Chinese urban system increasingly begins to resemble the western city in its embrace of market-driven land and planning allocation policies and urban boosterism, this process has not been accompanied by a parallel transformation of urban political society or governing elites. On the contrary, the economic reforms and high growth rates associated with the reforms of the 1980s have consolidated and extended the grip of state–party elites not only in globalising cities such as Shanghai and Beijing but right the way down to the county towns in the remoter provinces. Here rent-seeking opportunities for the allocation of prized development land (often the site of existing housing) are creating mutually beneficial alliances between local government officials and China's new entrepreneurs, while creating growing levels of urban inequality and urban poverty (Wu 2004).

Mexico

The Republic of Mexico, like China, was a one-party state for most of the twentieth century, but unlike China the federal state was closely connected to capitalist and agrarian interests, while the authoritarian apparatus of the Institutional Revolutionary Party (PRI) regime – although doggedly opposed to democratic reform and political pluralism – remained ideologically vague and inter-classist in terms of its constituency base, drawing support both from the urban middle class and the rural peasantry (Vellinga 1988: 154). In 2000 this picture began to change with the election of the right-of-centre National Action Party (PAN) followed in 2006 by a strong showing for both the PAN and the Revolutionary Democratic Party (PRD) which appeared to have ended the historic monopoly of the PRI for good.

The move towards a more pluralist polity in Mexico could already be detected at the municipal level by the late 1980s when a number of cities came under the control of opposition parties for the first time, particularly in the central and northern parts of the country. Under the PRI it was easy to see why political observers could identify the country as being representative of 'tight, centralized, top-down control, exercised mainly through the presidency and the corporatist structures established in the 1930s', even though in reality, as Cornelius has argued, 'a "Swiss-cheese" conception of political control in Mexico was more consistent with the historical evidence'(Cornelius 1999: 4).

Under the presidencies of Salinas and Zedillo from the late 1980s through to the 1990s it was clear that 'lower-echelon PRI leaders were no longer aligning themselves automatically with dictates from the center' and that under the 'new federalism' opposition party governments were able to control state and municipal governments amounting to some 55–60 per cent of Mexico's population (Cornelius 1999: 5–6). However, the new subnational political movements that are emerging to challenge the status quo whether from inside or outside the PRI are not necessarily pro-democratic. According to Jesús Silva-Herzog Márquez, 'with traditional [political] discipline having been shattered, political bosses, barons, and paramilitary squads are sprouting like mushrooms on the damp soil of the transition' (cited in Cornelius 1999: 10).

A notable exception is the Oaxacan city of Juchitàn where the indigenous Zapotec people's movement, the Coalition of Workers, Peasants and Students of the Isthmus (COCEI), succeeded in winning recognition from the federal government and administering welfare funds for a city of over 100,000 people, while promoting the Zapotec language and culture and mobilising the poor to demand a greater control of their economic futures. Having emerged from decades of persecution including state-orchestrated murders and military occupations, COCEI was able to seize the opportunity created by the less oppressive strategy of the federal government by engaging in local democratic elections and by stimulating grass-roots democratic forums within local churches, market squares, trade unions, municipal offices and family courtyards (Rubin 1999: 177). The political fruits of this successful mobilisation came in 1981 when an electoral alliance with the Oaxacan Communist Party brought COCEI to power in Juchitàn as the only leftist government in Mexico. The movement's leader, Mariano Santana, believed that the electoral victory heralded 'indigenous self-government' and an 'ethnic millennium'. With the red flag of COCEI flying at full mast in the municipal palace, 'after a long road with many sacrifices and obstacles', Santana declared, 'we had realized a Zapotec dream: to be the government' (cited in Rubin 1999: 186).

It soon became clear that both the state governor and the Interior Ministry had no intention of allowing COCEI to make a success of their new found administrative power, but internal rivalries within the State Federation of Chambers of Commerce and the fact that the governor lacked a political base in Oaxaca having come from Mexico City ruled out the possibility of a concerted attempt to remove COCEI. Instead, much of the municipal budget, which was controlled by the state governor, was cut off and state and federal loans and credits were denied. Only by organising protest marches from Juchitàn to Oaxaca was the state government pressurised into grudgingly and gradually restoring funding to the *ayuntamiento popular*, necessitating frequent similar protests (Rubin 1999: 187).

During its time in office in the 1980s, the COCEI was able to create new spaces of civic assembly, to insist that the Zapotec language be recognised and used in courts, police stations and in local administration; it promoted cultural festivals and events, helped to win better labour contracts for local workers and built alliances with rural workers and other municipalities throughout the isthmus (Rubin 1999: 189). The willingness of the Fox government to promote indigenous rights and scale back military occupations, though regarded as too limited by COCEI (UNHCR 2003), is, however, an indication of a new political geography in Mexico where subnational government can become a genuinely oppositional space and a focus for broader political mobilisation.

Conclusion

Understanding how cities are politically organised and what common denominators there might be across different national contexts remains a very challenging undertaking (see, for example, Bagnasco and Le Galès 2000). While it is relatively straightforward to point to different varieties of urban governance that can be found around the world, it is much more difficult to ascertain whether even liberal democratic city governments are beginning to converge on a single model of urban administration. Nevertheless, as our examination of London, China and Mexico's urban government has shown, one claim we can make with a good degree of confidence is that the political economy of cities is being rearticulated in terms of 'the competitive city'.

So far in this chapter we have compared and contrasted the emergence and operation of urban government in the pluralist, developmentalist and party–state traditions. The legal–bureaucratic apparatus of the local state plays a key role in all three contexts, but its sources of legitimacy are quite different. In the case of pluralist urban government, the open competition for public office affords credibility and a legitimacy to elected officials that allows them to make demands on higher-order governments and sometimes to resist (though not always successfully) the incorporation of local state interests within the routines of national and transnational policy making.

In developing states where Schumpeterian-type electoral competition is either non-existent, or recent and limited, as well as in more closed party–state systems, there is little to mediate the relationship of elites within cities and between urban, regional and national administrations – creating fertile opportunities for clientalism, corruption and more systemic forms of criminality as we found in Chapter 3. Even in these 'uneven geographies' of local state development it is nevertheless possible to identify a common trend in terms of a weakening of the monopoly of central state power, and a growing tension leading in some national contexts to a rebalancing of scalar power ensembles.

The universal impulse for this recalibration and reconfiguration of the urban is emerging 'from below' in the shape of popular demands for a more democratic and responsive government polity (as we noted in this and the previous chapter), and 'from above' in terms of the neo-liberalist deregulation and marketisation of urban governance as the new exigencies of capital demand an ever increasing role for the local state in eliminating barriers to accumulation. In the following chapter we shall investigate how the new territorialities of political and economic power are beginning to transform the 'new state spaces' of the contemporary city.

Further reading

In addition to the readings recommended in the previous chapter, those with an interest in the government of London and its role as a global financial centre will find *Regenerating London: Governance, sustainability and community in a global city* by Robert Imrie, Loretta Lees and Mike Raco (Imrie et al. 2009) a key text along with *Working Capital: Life and Labour in contemporary London* by Nick Buck and colleagues (Buck et al. 2002). There is no better guide to the inner workings of the multiple sites of London government than Tony Travers's *The Politics of London: Governing an ungovernable city* (Travers 2004).

Fulong Wu's edited collection, *Globalization and the Chinese City*, contains essays by leading scholars on the transformation of the Chinese city as a result of dramatic urbanisation and economic growth since the 1980s (Wu 2006). *Subnational Politics and Democratization in Mexico* by Cornelius, Eisenstadt and Hindley (Cornelius et al. 1999) remains a useful source of information on the development of urban and regional government, although there have been many important political developments in the intervening decade – not least the demise of Mexico's one-party state under the PRI. Paul Haber's *Power from Experience* (Haber 2006) charts the important role that organised movements of the urban poor played in helping to democratise Mexican politics and thereby helping to strengthen the country's nascent party system.

6 The confines of power

Cities, regions and states in a global perspective

> In the year 1238, the inhabitants of Gothia (Sweden) and Frise were prevented, by their fear of the Tartars, from sending, as usual, their ships to the herring fishery on the coast of England; and as there was no exportation, forty or fifty of these fish were sold for a shilling . . . It is whimsical enough, that the orders of a Mogul khan, who reigned on the borders of China, should have lowered the price of herrings in the English market.
>
> Edward Gibbon, *Decline and Fall of the Roman Empire*

> Scandalized by rampant corruption and unable to control their environment, the upper classes sought refuge in the villages on the periphery of the city, where market forces were pushing out the farmers. Public transportation and the automobile, which vastly increased the accessibility of the peripheral area, were drafted into the service of the fleeing upper and middle classes. Class relations soon manifested themselves in a new form: city–suburb relationships. At the same time businessmen shifted the focus of their political activities and attempted to run the city from the State House.
>
> James O'Connor, *The Fiscal Crisis of the State*

Introduction

Edward Gibbon's intriguing tale of the thirteenth-century herring trade has a particularly contemporary resonance. Fears of Asian prepotence, the dependence of entire industries on access to international markets, and the dramatic consequences that can entail when 'external shocks' ruin the ordered equilibrium of demand and supply fill the front pages of the press and feature in the news bulletins of the global media on an almost daily basis. Then, as now, statecraft, trade networks and military ambition gave rise to a world in which the lives of millions of individuals could be improved, disrupted or even destroyed by the rivalries of distant political elites, a sudden change in consumption habits, war or the vagaries of meteorological and climactic change. The more contemporary narrative offered by James O'Connor speaks to another form of territorial cleavage that has to do with the persistence of social divisions, the accretion of political power by a wealthy minority and the spatial secession of the haves from the have-nots, not only in advanced western societies, but wherever class-based societies are to be found.

In Europe, since the Treaty of Westphalia in 1648 the boundaries of the political became a defining characteristic of territorial sovereignty and their mutual recognition by other heads of state became both a necessary condition for the peaceful conduct of international relations and a frequent cause of conflict and war when these territorial divisions frustrated the ambitions of powerful expansionary states or the imagined communities of the unhappily ruled. The shift from a feudal concept of sovereignty that derives from the person and bloodline of the monarch to one in which territory itself is invested with a quasi-mystical and eternal 'national' character was key to establishing the particular and universal attributes of the nation-state from the latter half of the eighteenth century in Europe and in many of the colonies established by religious refugees in the 'new world'.

This elision of community with territory was a defining feature of the Protestant reformation and the political philosophy of the Enlightenment, where the notion of a territorially bounded self-government provided a counter to the universalism and 'divine despotism' of the Church of Rome.

With the establishment of territorially based states, a more refined and defined type of polity was made possible by virtue of the transcendent quality of 'the nation' – which eschewed the particular fatalisms of dynastic rise and fall. It is true that in England the Tudor state under Henry VIII and Elizabeth I succeeded by and large in instituting a single national confession, a quasi-autonomous rule of law and a small but effective military–bureaucratic apparatus – but the personality and vulnerability of the monarch and his or her descendants to rival claims on the throne made the business of rule tenuous, contingent and prey to territorial reconfiguration through war and marriage. In France, the mercantilist statecraft of Louis XIV's finance minister, Jean-Baptiste Colbert, did much to advance the modernisation of the state apparatus – but the collapse of the House of Bourbon under Louis XVI revealed the vulnerability of dynastic absolutism to the claims of a burgeoning territorial nationalism (Taylor 2003: 105).

As we saw in Chapter 2, cities played a key role in this territorialisation of power and as agents of national integration. However, the trajectory has not always been a linear one – even in the European context where religious, economic and cultural affinities ought to have made for a greater consistency of urban form than in other continents. With the advent of nation-building and an increasingly urbanised population in eighteenth- and nineteenth-century Europe, North America and to some extent Latin America there emerged more metrocentric forms of governance where power shifted from a localised feudal lordship dependent predominantly on landed capital to a money-based economy, which along with the financial institutions necessary to provide currency and credit also required effective forms of communication and administration and a large degree of labour and capital mobility. The city, as we have observed, uniquely provided the spatial milieu within which this synekism could take place, but the degree to which individual cities were able to exploit this power varied considerably.

Therefore if urbanisation is at the heart of the transformation of the state into what Giddens has called a 'power container' (Giddens 1985), it is important, as Friedmann and Wulff remind us, to appreciate that urbanisation has two distinct meanings. The first is associated with 'the concentration of formerly dispersed populations that are primarily engaged in farming in a small number of settlements whose principal economic activities are in the services, trades and manufactures', while the second refers to 'urban modes of production, living, and thinking originating in these centres and spreading from these to outlying towns and rural populations' (Friedmann and Wulff 1976: 4). The authors remind us at the same time of the necessity of approaching the study of regionalisation and urbanisation from a historical perspective, of the need to use a macro-lens when considering spatial systems that are organised around a 'core' and a 'periphery', while also arguing in favour of micro-studies that are attuned to the fine-grained nature of 'urban morphology, social organization, and mobility; the urban economy, and urban politics and social control' (ibid. 5).

This fuller understanding of urbanisation allows us to think about cities and regions less as mere containers of populations and economic activity and more as a dynamic spatial complex involved in the transformation not only of its own territory and landscape, but as a powerful agent of national and transnational restructuring and rescaling. The particular ways in which states, regions and cities engage in rescaling and restructuring are explored in the section that follows, but any critical understanding of the changing boundaries of state power must also ask the following question. For whom and for what purpose are territorialised polities undergoing these particular changes? The refashioning of state forms and institutions is also necessitated by the 'failure' or crisis of pre-existing modes of governance. Thus a third analytical perspective is proposed – that of 'remediation' which allows for a more dynamic understanding of the relationship between state forms and the variety of social systems within which they are embedded.

Adapting the definition provided by Richard Grusin in relation to new media technologies we can say that remediation is the double logic according to which the state (particularly but not exclusively the neo-liberal state) refashions prior state forms (Grusin 2004: 17). It is a double logic because at the same time as new state forms are refashioning pre-existing states, older state forms are adopting the modalities of new state configurations in order to 're-invent' themselves, thereby extending their governing capacity and longevity. At the same time, particular modes or regimes of capitalist accumulation impose demands on the strategic and functional constitution of governmental systems, inducing what Joseph Schumpeter has termed a process of 'creative destruction' at the institutional as well as the societal level (Schumpeter 1987).

Rescaling, restructuring or remediating? States, regions and cities

There has been a resurgence of interest in the spatial characteristics of states in recent years that Brenner and colleagues attribute to the increasing influence of Henri Lefebvre's work on scholars investigating the relationship between state and power, particularly following the publication in English of Lefebvre's landmark book *The Production of Space* (Lefebvre 1991), the growing significance of 'globalisation' as an analytical category for the transformation of states and economies since the 1970s, the crisis of the Keynesian welfare state leading to a decentralisation and deterritorialisation of government, and finally the emergence of new localisms or new regionalisms associated with forms of governance built around the 'growth machines' or urban/regional regimes that have as their primary goal economic regeneration and community building (Brenner et al. 2003: 3–4), which we discussed in the previous chapter.

We can therefore point to two separate but related factors that help to explain 'the spatial turn' in state theory. One relates to an epistemological refocusing on space and territory in the critical social sciences as not merely the place where conflicts and realignments of power happen but rather as principal agents of change in their own right. The second factor relates to institutional and organisational transformations within and between states in the last three to four decades that have been attributed to what O'Connor (O'Connor 1973) has called 'the fiscal crisis of the state', and what Peck and Tickell refer to as the 'rolling back' of direct state intervention in a range of policy areas at different levels of scale and the 'rolling out' of neo-liberal strategies aimed at the marketisation of all forms of social and economic relations including the public, the private and the voluntary sectors (Peck and Tickell 2002).

Even if we accept that these transformations are indeed occurring, it soon becomes obvious that the territorial claims of the globalisation advocates are rather different from those of the post-Keynesian welfare state and neo-liberal analysts. The geography of neo-liberal post-welfare state restructuring is undeniably centred on the North Atlantic – with the United Kingdom and the United States forming the twin pillars of the colossus that Jessop (Jessop 2000) has described as the Schumpeterian Postnational Workfare State – whereas by its very nature, globalisation is systemic and worldwide in its reach and impact. It follows then that many of the state systems affected by processes of globalisation have either not developed as Keynesian-type welfare states in the first place (the great majority of national states), while others that have (especially in the Nordic countries and continental western Europe) cannot be counted among the first wave of neo-liberal state transformers, though many are caught up in a less violent second wave in which the market is playing a more significant if not yet dominant role in state–society relations. As Peck and Tickell write, a more nuanced discussion of neo-liberalisation would therefore 'need to take account of the ways in which ideologies of neoliberalism are themselves produced and reproduced through institutional forms and political action, since "actually existing" neoliberalisms are always (in some way or another) hybrid or composite structures' (Peck and Tickell 2002: 383).

However, before launching into a discussion of the refashioning of state spaces under neo-liberalism it is first necessary to account for the dominant state forms within which city and regional governments have developed since the early twentieth century during the era of large-scale mass manufacture known as Fordism.

Fordism, post-Fordism and neo-liberalism in urban and regional governance

Since the 1970s a body of literature has developed around the role of the state in the capitalist economy which has become associated with its chief analytical focus – the regulation of civil society (and especially economic activity) by state actors on an international, national, regional and local scale. Regulation theory draws heavily on Marxist analysis and gives the state a central role in the direction of the capitalist accumulation regime. Lipietz refers to the phrase in *The German Ideology* of Marx and Engels that defines the state as 'the apparatus by which society equips itself in order that the different groups of which it is constituted do not exhaust themselves in a war without end' (Lipietz 2003: 241). In other words, in a world of a war of all against all, the state's function is to secure not only peace and security, but to ensure the conditions by which the dominant mode of production is able to reproduce itself and protect its interests. Consequently regulationists have sought to periodise the development of capitalism in the twentieth century around the emergence and growth of Fordism as both a dominant mode of production and a state-directed regime of accumulation that structures social and economic life and organises labour and capital in particular spatial distributions.

The British state theorist Bob Jessop identifies four different meanings of the term Fordism: labour process, regime of accumulation, mode of regulation and mode of societalisation (Jessop 1992 in Goodwin and Painter 1996: 640). In economics, however, Fordism is most often referred to as an automated assembly-based production system that allows the manufacture of goods (such as automobiles) in higher volumes by the use of specialist automated tools and constant innovation and improvement of the product in terms of quality and reliability. The spatial concentration of production under one roof from the input of raw materials to the rolling-out of the finished product (known as vertical integration) requires a large, skilled and proximate workforce, while the site of manufacture has to provide good access to regional, national and international supply and distribution channels.

> As a mode of regulation (MOR) Fordism is associated with (1) a wage relation in which wages are indexed to productivity growth and inflation; (2) a key role for the state in managing demand; and (3) state policies which help to generalize mass-consumption norms. The state operates demand management policies in the fiscal sphere and underwrites a minimum level of working-class consumption to complete the virtuous circle.
>
> (Goodwin and Painter 1996: 640)

These macro-economic regulatory features are also typical of Keynesianism where the state plays a central role in ensuring the conditions for economic growth by regulating market disequilibria through a variety of direct and indirect fiscal and monetary interventions. Fordism shares another attribute of Keynesianism which differs from laissez-faire economics in terms of its concern with ensuring a measure of social well-being and an obligation to the workforce that goes beyond the cash nexus of the wage contract. Hence 'welfare capitalism' in America developed as a mode of societalisation through the provision of high relative wage levels (Ford notoriously doubled the basic wage at his Detroit manufacturing plant in 1914 to $5 an hour, much to the chagrin of the *New York Times* and local business leaders), a reduced working week, and a range of social benefits. Henry Ford's aim was to secure a loyal and contented workforce that would reject the appeals of the labour unions and give more freely of their effort and creativity because employees would feel that the company valued them as individuals and not merely as commodities to be acquired for the cheapest possible price.

Ford was not the first to appreciate the advantages of winning the hearts and minds of the working class through welfarism and higher wages. Bismarck's introduction of a social welfare system in Germany in the 1880s was not only a ploy to demobilise an increasingly active and growing labour movement and to wrong-foot the German Social Democratic party, it was also a means of replacing complex industrial injury laws with a state-sponsored system that displaced the cost of workers' compensation from companies onto German tax payers (Hennock 2007). In other words, state welfarism's primary purpose was the socialisation of entrepreneurial risk as a stimulus to enhanced capitalist accumulation.

Even in the United States where the federal government played a more minor role in the planning and support of regional industrial agglomerations – city and state-wide governments were keenly aware of the need to orient their infrastructural investments and economic support to big capital, even at the risk of alienating support from their traditional constituencies such as farmers, merchants and small business owners. However, as Theda Skocpol notes, unlike in Europe, 'U.S. political institutions have never allowed the possibility for [a] comprehensive regime of public social provision to emerge' (Skocpol 1995: 33). The strict separation in American social policy between 'social security', which is a contributory unemployment and pension benefit paid for and to those in long-term work, and 'welfare', which is regarded by government and public opinion as a hand out to the undeserving poor, has antecedents in the utilitarian poor law reforms that helped to shape British social policy in the nineteenth and early twentieth centuries.

Thus it is not incidental that Fordist 'welfare capitalism' is associated first and foremost with the disciplined production and reproduction of an optimal supply of labour, which requires the active participation of the state through the provision of education, housing, land and transportation in order to spatially concentrate human resources within the confines of the Fordist productive system. This is why in his book, *The Urban Question*, Manuel Castells (Castells 1977) argued that in essence the modern city had become a vast capitalist factory where the hierarchies found on the shopfloor were reproduced through the control of public space and social amenities by the middle classes, by forcing the proletariat to consume the products of their own alienated production and to pay for the reproduction of their own labour in the form of rent to bourgeois rentiers or the capitalist state. In other words, the Fordist city was nothing other than the expression of the spatial logic of a class-divided capitalist economy built on mass manufacture and consumption.

Along with a number of other radical sociologists, Castells gained insight into the conditions of the urban working class as a result of the social protests that enveloped many European cities during and after the events of 1968 (see Chapter 4). These contestations were concentrated within the major productive centres of the western economy, but similar convulsions were occurring in Latin America, southern Africa and Asia. The crisis of Atlantic Fordism (the Rhineland model being an important exception) was resulting in a declining rate of profit, over-capacity as a result of falling demand, and declining competitiveness as newly emergent economies such as Japan replaced the Fordist manufacturing model with a differentiated 'just-in-time' mass-production system based on the Toyota model (Fujita and Child Hill 1995).

Ironically, even at the point when Atlantic Fordism was entering a period of terminal crisis in the early 1970s, the local state was restructuring itself on firmly Fordist principles. As Goodwin and Painter note in the context of the United Kingdom, the emergent local Fordist state could be recognised through the deployment of elected local government as a key apparatus of the Keynesian welfare state, by the provision of social wage subsidies in the form of local authority administered housing programmes, the provision of services that were essential for the purposes of regulation (such as licensing and environmental health) but which the market could not collectively provide, the management and financing of infrastructure and planning, and also in terms of the local state's increasingly centralised, hierarchical and managerial organisational form (Goodwin and Painter 1996, 461–2).

It was not that the role of the local state as an agent of social welfare was novel – poor relief, for example, had been a function of local government as far back as the Tudor period in England (Slack 1990); rather it was the simultaneous transformation of the local state into a major economic actor in its right (both as a landowner, rentier, employer and investor), as a universal welfare provider and increasingly as the chief agent of territorialised accumulation strategies, resulting in increased institutional complexity and the need for novel spatial divisions of administration as the new era of corporate local government sought to map itself onto the disappearing contours of territorial Fordism.

From city-regions to global cities

It is a puzzle why, as Jane Jacobs reminds us, that cities such as Tokyo, London, Paris and Milan have been capable of generating metropolitan regions whereas Glasgow, Edinburgh, Marseille and Naples have not (Jacobs 1985: 45–6). What induces this metastising tendency in certain large settlements and not in others cannot be attributed to a single factor alone, but it has to do with a combination of determinants including geography (core cities are much more likely to regionalise than peripheral cities), the relative population density of the rural hinterland, spatial and other intra-mural limits to growth, the availability of surplus investment capacity, the capacity and extent of transport and infrastructural networks and so forth.

All of these features helped to establish the Tokaido Corridor (also known as the Taiheiyō Belt) in Japan as one of the most significant of the world's megalopolises. Stretching some 1,200 km from the city of Kobe in the west to Tokyo's Narita airport in the east, this highly urbanised and industrialised coastal strip includes a population of approximately 83 million or just over 65 per cent of Japan's total population. The Greater Tokyo region alone accounts for 35 million people who are crowded on to a landmass that amounts to just 0.6 per cent of the national territory. The Greater Osaka region accounts for 17 million and the Chukyo metropolitan area (at the centre of which lies the city of Nagoya) is home to some 8 million of the Tokaido Corridor residents. According to Yamamoto (Yamamoto 1987) the post-war history of Japan's urban development can be explained by 'a ceaseless destruction of regional subsystems combined with a reorganization of the country as a single spatial unit'. This concentration of migratory flows, transport infrastructure and economic activity has seen the considerable growth of the Tokyo and Osaka metropolitan areas together with 'the integration of the urban network under the "hegemony" of Tokyo, the capital'. As Yamaguchi writes, '[i]n this sense, Tokyo is not only the nation's pre-eminent city but also the world's largest banking centre and the world's third largest corporate centre following New York and London' (Yamaguchi 1988).

A powerful coalition of state-led development and major industrial corporations, together with a rapid transformation in the Japanese economy from an agrarian to a manufacturing economy in the first decades after the Second World War ensured that the Tokyo–Osaka growth poles would eventually absorb the populations of the smaller regional towns and the rural hinterland. In a uniquely Japanese version of developmental Fordism, which gave little room for alternative development models at the municipal or prefectoral level, the urbanisation of the Tokaido Corridor reflected the continuing importance of national–regional and sub-regional hierarchies based on Tokyo's increasing importance as a global financial centre and as a 'command and control' centre for Japan's highly successful transnational corporations.

If we move our focus to the northern hemisphere it is clear that the interconnections between city–regional clusters produce similar spatial agglomeration patterns to those found in Japan. For example, the economic mega-region identified by the French geographer Roger Brunet known as the 'blue banana'

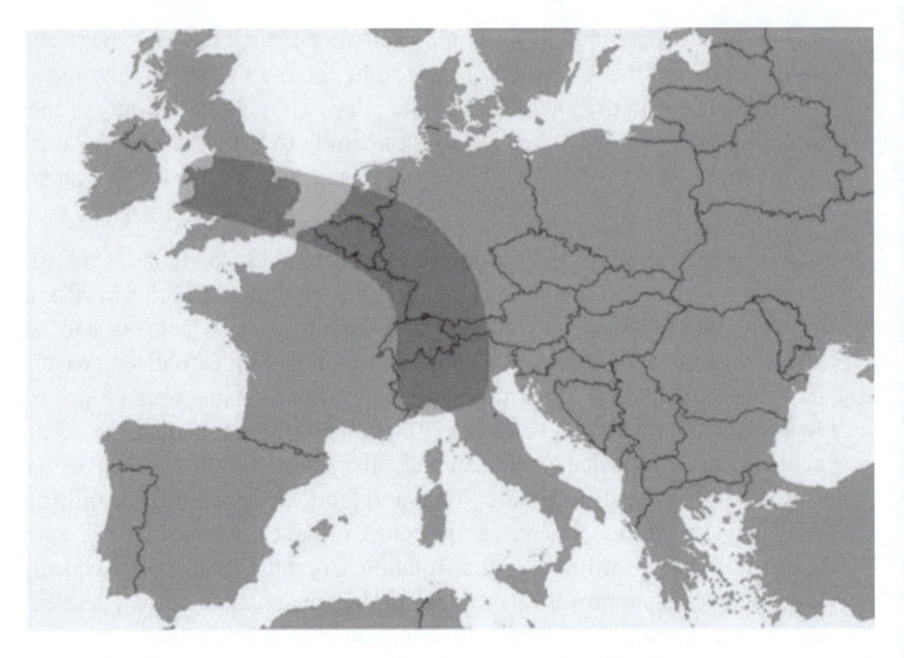

Figure 11. The European mega-region known as the 'blue banana'.

stretches from Manchester in the north-west of England and curves down through Greater London forming a crescent running from Paris in the west to Frankfurt in the east down through Bavaria and south as far as Milan and Bologna in northern Italy (see Figure 11). To this map one should add the 'golden crescent' of the northern Mediterranean between Barcelona and Genoa, an Atlantic arc that stretches from Galicia to Merseyside and a Baltic arc that includes the old Hanseatic port cities of Hamburg and Riga (Brunet 2002). Brunet and his colleagues at the French regional planning agency (DATAR) wanted to draw attention to the continental dimension of Europe's regional economic structures and to eschew the narrow nationalism that had dominated regional planning policies hitherto.

Although the 'blue banana' has been subject to a good deal of criticism for failing to capture the true spatial distribution and extent of the 'hot' economic zones in western Europe, there has been a widespread acceptance of the increasing integration of transnational regional economies – a trend that has been exacerbated and accelerated by the expansion of the European Union to the Baltic and the Caucasus – leading Brenner to conclude that these extensive urban–regional corridors are 'a powerful expression of the centripetal, polarizing forces that have been unleashed since the crisis of North Atlantic Fordism' (Brenner 2004: 188).

Global cities and globalising cities

The return of the city as an object of analysis in the ordering (or re-ordering) of the world system marks a shift from the Westphalian model of multiple but equal territorial sovereignties to a more complex, multi-scalar and 'uneven' global space in which it is more relevant to talk about investment and migration flows, the density and extent of productive capacities and consumer markets. Here the network – whether it takes the form of pervasive communication technologies such as the Internet – or the collaborative associations that bring together a variety of actors in pursuit of a common goal (be it profit, policy delivery or the social good) is a ubiquitous binding that permits these relays and connections between spaces and scales to assume a self-organising function.

As we noted in Chapter 2, at the centre of every spatial territorial network is a town or a city that at its most basic level operates as a node, a space where communication paths converge and where goods, money and labour are exchanged. Max Weber recognised in the urban market the core function and utility of the city whose monopoly of communication, exchange and finance ensured its leading role in the development of capitalist modernity. In a similar vein, Roger Keil argues that globalisation 'makes states', but these states are different in many ways from those that existed at the height of Fordism from the 1920s to the early 1970s. In particular, if we think of city government as a 'local state' it is important to immediately qualify this description by pointing to the fact that for many of the world's largest cities their economic, cultural and social compass is anything but local – and their governance operations, while including the local, also involve in many cases a regional, national and indeed international function. From this perspective it is clear that cities like Brussels, New York and London are sites of multi-scalar governance that are as much the product of globalised and globalising power networks and urban spaces as they are sub-national polities (Keil 2003: 278–95).

The literature on the functional features of global cities and why these are indispensable to the operation of the international economy is now very extensive and well known (see, for example, Friedmann and Wolff 1982; Sassen 2000; Sassen 2001; Taylor et al. 2003; Taylor 2004). There are many ranking options that one could adopt and weights that might be given to different global city status characteristics – but these are little help in explaining what power such cities possess, of what such power consists and how it is deployed at different territorial scales. The political significance of Geneva because of its importance to the international diplomatic and NGO community is significantly greater than Zürich's, for example, but the position would be reversed if we considered power from an economic and financial perspective because of the high concentration of international banks and insurance companies that are concentrated in Zürich.

The growing importance and variety of global cities is happening at a time when national regulatory/accumulation strategies have proved inadequate to deal with the crisis of global Fordism and should be seen as part of, 'the emergence of urban-regional economic networks', and therefore as 'a complementary process of local economic agglomeration nested within a wider regional and national system of global integration' (Parker 2004: 115). Globally important cities are therefore at the centre not only of world networks of goods, services, labour and capital – they are the dominant urban agglomeration within their regional economic systems and a vital source of employment, housing, transportation, education and much else besides for a considerable local population, most of whom will firmly remain within the 'micro scale' even if many will have been born in a different country. This 'glocal' (Swyngedouw 1997; Brenner 1998; Swyngedouw and Baeten 2001) aspect of strategically important cities helps us to understand the hybrid and multi-scalar nature of globalising urbanism and why the contextual particularities of political cultures and traditions and institutional practices and structures require us to qualify the notion that a super tier of global megalopolises is in the vanguard of a 'run away world' (Giddens 2003).

Indeed, as Goran Therborn and K.C. Ho remind us in an editorial for a special journal issue on the south-east Asian metropolis:

> [a]longside the currency of economic power are violence, state command, democracy or dictatorship, judicial and legislative systems, mass manipulation or mass mobilization, and power needs to [be] legitimized and symbolized. Before being locations of business headquarters, cities are built environments pulsating through concentrations of people, people of acquiescence or rebellion, of poverty and hopes for something better, as well as people of wealth and power.
>
> (Therborn and Ho 2009: 54)

Therefore if we wish to understand the ways in which the uneven geographies of wealth and exploitation are unfolding across the majority urban world, we need to pay more attention – as Jenny Robinson (Robinson 2006) argues in the context of urban South Africa – to the many 'ordinary cities' in which hundreds of millions of the world's population now reside.

The governance of new state spaces

The term most associated with the rescaling of metropolitan–regional spaces is 'new regionalism', a concept that is more familiar to students of international relations or international trade than to the study of local politics. Urban new regionalism's origins can be traced to a series of studies that came out of the United States in the 1980s and 1990s that attempted to understand how America's larger urban centres were responding to the challenges of an increasingly global competition for investment, qualified and skilled workers and entrepreneurial talent. The biggest obstacles facing the longer-established American cities was the absence of a regional-scale metropolitan government with the powers and resources to effectively plan and manage an integrated city–region. Instead, cities such as New York or Los Angeles had to coordinate between dozens of legally autonomous municipalities, school boards, transit and port authorities, and state and federal agencies – while the majority of those employed in the metropolitan core did not contribute their taxed income to the upkeep of the city.

The dire financial situation that the major US cities found themselves in by the 1980s and the lack of effective powers to coordinate economic growth led a number of state governors and big city mayors to call for 'a new regionalism' that would facilitate cooperation between metropolitan and suburban local government and introduce a degree of coordinated planning to the economic strategies of the different regional actors. According to Brenner the motivation for this search for a scalar shift in urban governance was a consequence of 'the new forms of sociospatial polarization and uneven geographical development that have been crystallizing in US city-regions under conditions of postfordist urban restructuring and neoliberal (national and local) state retrenchment' (Brenner 2002: 3).

The political impetus for these initiatives has mostly come from state governments that have seized on the need make their city–regions more economically competitive by

> coordinat[ing] the activities of competing municipalities in a metropolitan region according to shared priorities for regional growth . . . establish[ing] a regional framework in which local planning policies, infrastructural investments and other aspects of inter-local governance may be coordinated . . . pool[ing] fiscal resources at a regional level . . . and . . . limit[ing] environmental destruction through the establishment of compulsory region-wide land-use planning.
>
> (Brenner 2002: 13)

Whereas in previous incarnations of regional integration movements the aim was to mitigate the effects of place-based inequality (Swanstrom 1996), the objective during and after the Reagan era was to promote economic growth and prosperity. It is true that among pro-regionalist liberal reformers grace notes in praise of building strong citywide communities and tackling urban decay could be heard above the growth-centred chorus of business and political elites, but in the United States there has been little enthusiasm for the amalgamation of city and suburban governments in the form of metro-regional authorities or nationally coordinated regionalisation strategies more typically associated with Europe and certain Canadian provinces such as Ontario (MacLeod 2001; Brenner 2002; Keil and Kipfer 2002). Indeed only two of the most populous US cities in the 1960s (Milwaukee and Chicago) underwent a process of annexation leading to an expanded metropolitan government region in the period after 1945 (Goldsmith 2001: 327).

Ontario has a resident population of 12.5 million, but nearly half that total (5.5 million) live in the city–regional agglomeration of the Greater Toronto Area (equivalent to 8 per cent of Canada's population and 30 per cent of all recent immigrants to Canada). The population of the City of Toronto at just over 2.5 million in 2006 represented 45 per cent of the GTA. Toronto has one of the highest urban growth rates in North America, and its economy accounts for 11 per cent of the total Canadian GDP. After a period of local government reorganisation in the mid-1960s a second phase of 'top-down' restructuring began in the late 1990s.

According to Leibovitz, the pressure for a Canadian version of 'new regionalism' in Toronto came from two different directions. One was the result of the expanding welfare state, which encouraged the provincial and federal government to increase their social regulatory role and to ensure greater equity and consistency of service delivery at the municipal level. This was in line with the type of Fordist-corporatist restructuring that occurred in Britain following the Redcliffe–Maude report in the early 1970s (Wolman 1995: 139–40). The second impulse towards the upscaling and centralisation of metropolitan government occurred in the 1990s, and this time the rationale was the need to respond to the fiscal crisis of the federal, regional and to an extent local state (Leibovitz 1999: 202–3; see also Hackworth 2007).

Keil and Kipfer put the case more strongly, arguing that 'the hard times of the 1990s and municipal restructuring signal a new phase in Toronto's planning history'. Having closely examined the restructuring of governance in the newly amalgamated City of Toronto, they conclude 'that Toronto is being moulded into a "competitive city" defined by a complex of class alliances and political coalitions, neoliberal planning and economic policies, multicultural "diversity management," and revanchist law-and-order campaigns' (Keil and Kipfer 2002:

229). But as one of Toronto's most celebrated residents, Jane Jacobs, observed in one of her last books, if the ostensible aim of the reform was to reduce costs, the new administration turned out to be 'expensive for many reasons, some unavoidable, some irresponsible, some apparently corrupt, all of them deplorable' (Jacobs 2004: 119). By the end of the 1990s, Toronto re-emerged as 'a progressive city with a neoliberal twist' with the election of Jane Jacobs's favoured candidate, the reformist mayor David Miller, who pledged to 'clean up' Toronto's streets, air and City Hall (Boudreau et al. 2009: 199–202).

The new urban entrepreneurialism

The recent history of Toronto is emblematic of a global trend that the Organisation for Economic Co-operation and Development celebrated in its report, *Competitive Cities: A new entrepreneurial paradigm* (OECD 2007). Commenting on the shift from a predominantly managerialist/social Fordist mode of urban government to a 'new public management'-based market-oriented model, the OECD prefaces its report by claiming that

> competitive cities will have to shift from the managerial mode of policy making, concentrating on managing spatial needs by land-use control and infrastructure provision to the 'new urban entrepreneurialism', which focuses on creating new industry and jobs by attracting an economically active population into inner areas and regenerating economic infrastructures.
>
> (OECD 2007)

The OECD is clear that economic globalisation and increasing competition between cities around the world has led to a shift from managerialism 'which is primarily concerned with effective provision of social welfare services to citizens . . . to entrepreneurialism, which is strongly characterised by pro-economic-growth strategic approaches, risk-taking, innovation and an orientation toward the private sector' (OECD 2006: 7). Urban entrepreneurialism is defined in the following ways:

- Fostering and encouraging new forms of local economic development in a 'pro-active' rather than 'passive' manner (e.g. by the use of 'positive planning' instruments such as special economic development zones)
- Making full use of the market to achieve public sector goals and much less use of public intervention
- A strong pro-private, pro-business agenda involving strategic alliances such as private–public partnerships
- A willingness to engage in risk-taking, innovation, profit motivation and other growth oriented incentives more typically associated with the private sector.

Cities have thus become adept at playing up their strengths and playing down their weaknesses in a global 'pick me' fight for prestige events (such as Olympiads, Expos, inter-governmental meetings, cultural festivals), strategically important institutions (such as the European Central Bank), high-profile sports and artistic personnel and venues (baseball and ice hockey teams, Guggenheim museums, concert orchestras), global corporate headquarters, international airport hubs, 'smart growth' pioneers and so on.

As Hall writes, 'Mega-events are . . . one of the means by which places seek to become "sticky" . . . – that is attract and retain mobile capital and people – through place enhancement and regeneration and the promotion of selective place information' (Hall 2006: 59). In the view of a major international NGO involved in campaigning against the displacement of urban populations, '[t]he staging of a mega-event is typically motivated by three key concerns:

(1) putting the city 'on the world map' (increasing tourism);
(2) boosting economic investment in the city and attracting capital (for improving urban infrastructure and redevelopment); and
(3) 'reimagining' the city'.

(Centre on Housing Rights and Evictions 2007)

The focus on the growth/regeneration attraction of mega-events, which has an obvious lure for real-estate developers, the construction, leisure, tourism and entertainment industries, as well as local political leaders and officials for whom the hosting of a successful major event is likely to be the high point of their careers,[30] often obscures the hidden costs of making the city 'fit for global consumption' (Surborg et al. 2008).

According to the Centre on Housing Rights and Evictions (COHRE), the Olympics have been responsible for displacing more than 2 million people over the last 20 years in the various parts of the world where the summer and winter events have been held – a figure that includes between 1.25 and 1.5 million people alone in China in the run-up to the Beijing games.

In an important sense, spectacular urban entrepreneurialism, all the way from high-profile globally networked and consumed events such as Expos and Olympics to local 'community pride' celebrations, is about valorising and promoting urban spaces as places for the production and consumption of 'experiences' (or spectacle) rather than material goods (although of course material commodities and infrastructure are essential to making the spectacle possible), and de-valorising others – especially spaces where the 'visible poor' such as the homeless or migrants congregate. As such, spectacular urban entrepreneurialism has close affinities with a significant new field in growth machine politics that we might term 'creative boosterism'.

The creative city

The type of city marketing that is associated with mega-events is but one aspect of a successful 'competitive city' strategy which since the publication of Richard Florida's *The Rise of the Creative Class* in 2002, has placed considerable emphasis on attracting the new 'creative class' (which Florida estimates to amount to some 38 million individuals or 30 per cent of the workforce in the United States) to 'creative centres' where they feel comfortable to live and work. Florida sees the major change in regional development strategy over the past two decades as being a shift away from locating the factors of production in the most optimum space (which often lay outside traditional urban centres in new 'edge cities') to a reinvestment in 'place and community' (Florida 2003: 4).

City leaders that want to catch the wave of the global knowledge economy should not, however, be seeking to reproduce the world of the 1950s, according to Florida, where 'strong ties existed between families and friends', leading to 'close neighbourhoods, and those attributes that come along with such communities, such as civic clubs and vibrant electoral politics' (Florida 2003: 6). Instead the accent should be on encouraging creative human capital which is characterised by weak or 'quasi autonomous' social ties. The new creative class seek places that are 'innovative, diverse and tolerant' and their working lives are directed towards creating 'meaningful new forms'. Creative-class professionals, it is claimed, are not attracted to traditional features of major population centres such as shopping malls, sports stadiums and tourism and entertainment districts, but instead they look for openness to diversity, support for the arts and small-scale enterprise, and 'the opportunity to validate their identities as creative people' (Florida 2003: 9).

The creative-city idea has also been championed by UNESCO (the United Nations Educational, Scientific and Cultural Organization), which launched a 'Creative Cities Network' in 2004 involving a number of cities around the world that were prepared to support and sustain so-called 'creative clusters'. UNESCO's network has focused on cities, 'because they are increasingly playing a vital role in harnessing creativity for economic and social development. Cities harbour the entire range of cultural actors throughout the creative industry chain, from the creative act to production and distribution'. Some of the first UNESCO creative cities include Aswan in Egypt, Buenos Aires in Argentina, Santa Fe in New Mexico, Popayan in Colombia and Edinburgh, Scotland – each of which has its own cultural 'unique selling point' (e.g. folk art or gastronomy).[31]

According to their advocates, 'creative cities have managed to nurture a remarkably dynamic relationship between cultural actors and creativity, generating conditions where a city's "creative buzz" attracts more cultural actors, which in turn adds to a city's creative buzz. This virtuous cycle of clustering and creativity that is shaping the foundation of creative cities is also perpetuating the evolution of the "new economy"'.[32]

Florida talks of the '3Ts' of economic growth as being fundamental to the creative city phenomenon: these are technology, talent and tolerance. Technology relates to the exploitation of human knowledge, while talent is about cities bringing together and augmenting human capital – which, following the economist Robert Lucas, Florida refers to as 'Jane Jacobs externalities'. 'In this sense', Florida continues, 'urbanization is a key element of innovation and productivity growth'. The third T relates to tolerance, which is seen as 'the key factor in enabling places to mobilize and attract technology and talent'. According to this aspect of the creative-class thesis there is 'a strong connection between successful technology – and talent harnessing places and places that are open to immigrants, artists, gays, and racial integration . . . Such places gain an economic advantage in both harnessing the creative capabilities of a broader range of their own people and in capturing a disproportionate share of the flow' (Florida 2005: 6–7).

A fourth T has been added by the Creative Class Group (responsible for the public policy and commercial dissemination of Florida's ideas) – 'territorial assets', or locational amenities that would include everything from a network of cycle paths to world-class concert orchestras. Each iteration of the creative-class recipe for economic growth is premised on the assumption of high degrees of geographical mobility and human capital allied to urban knowledge resources that will allow one to find an ideal home in the city that best fits one's tastes, preferences and socio-demographic make-up (Florida 2008). This notion of 'human enterprise' as optimal consumption and vocational place seeking has many echoes with the new forms of 'extensive' habitus and social identity typical of the affluent, mobile middle classes that we explore in Chapter 7.

Largely as a consequence of these myriad individual location decisions, as Florida himself admits, '[t]he creative economy is giving rise to pronounced political and social polarization – a demographic sorting process that separates people by economic position, cultural outlook, and political orientation. This *big sort* is further aggravated by the perception among many that key elements of the creative class are arrogant, hedonistic, and self-indulgent' (Florida 2005: 172). This is a 'spiky' urban world in which human-capital-rich creative oases are often surrounded by neglected inner cities and declining middle-class suburbs.[33] As such the intensifying socio-economic segregation of urban populations is the human consequence of the systematic abandonment of the state's former welfarist and regulationist function under the social compact of Atlantic Fordism, which has been replaced by more coercive and revanchist governmentalities that both eschew and devalorise the democratic and republican values of the urban civic community (Low and Smith 2006; Wacquant 2008).

Conclusion

As the World Bank informs us:

> Urban areas sharing large regional markets (border zones and port cities, such as those surrounding the South China Sea) are becoming closely networked, sometimes developing interdependencies across national boundaries that are as close as, or even closer than, those with their own hinterland. By reducing the traditional market position of some cities and fostering the growth of others with different locational and production advantages, the liberalization of trade and financial flows is contributing to large spatial shifts of population and output. These changes imply that now more than ever, cities need to provide solid public services and a business-friendly environment to retain their traditional firms or to attract new ones, domestic or foreign.
>
> (World Bank 2000 in Olds and Yeung 2004: 499)

This is a good illustration of the process through which globalisation leads to a convergence of different 'path dependent' urban regimes onto the single economic field of the competition state. It is still possible to identify the distinctive characteristics of developmental city-states such as Singapore, which are very different to, for example, Manchester or Chicago, but each of these cities is bidding to secure significant amounts of international investment, to retain and increase its firms' share of the global market, to attract tax-paying companies and employees and high-spending visitors, and to mitigate the costs and impacts of welfare dependency while simultaneously ensuring a constant supply of 'flexible' unskilled labour to satisfy the increasing demand for individualised consumption and mobility.

As Ferman writes, '[w]ith an ideology and political culture that promote a diminished role for government (i.e. except as supporter of private sector activity), that assume a definition of economic development policy limited almost exclusively to bricks and mortar, and that employ a narrow set of policy evaluation criteria, it is hardly surprising that urban policy agendas are still dominated by a growth machine mentality' (Ferman 1996: 147).

However, it would be a mistake, as Kirby and Abu-Rass comment, to see entrepreneurial urbanism as a product of enterprise alone. Indeed, even in the USA, at every scale of the polity from the federal to the municipal 'urban growth has been stimulated or constrained by the decisions made by [state] agencies, both civilian and military' (Kirby and Abu-Rass 1999: 224). What is significant in the turn towards an entrepreneurial mode of urban governance is the privileging of market values and activities over other more traditional local government functions and priorities (such as the provision of local welfare services and the provision of locally consumed amenities). This neo-liberal urban imagineering sets great store by the visual appearance of public spaces as either enhancing or detracting from the 'marketability' of the city as a space of collective private consumption and investment (Coleman 2005).

The enthusiasm for the 'creative cities' moniker among contemporary urban boosters serves to highlight the fetish for global brand status but it tells us little about the real status of cities within the hierarchy of urban power that we discussed in the previous chapter. If the British 'bohemia index' can place Manchester first, and Leicester joint second with London it would be immediately obvious to a British sociologist that the 'diversity index' is over-determining the independent variables that the index should really be trying isolate, such as the strength of the cultural industries and the new economy.[34] Equally, the desperate attempts by city mayors and their national governments to host sporting and cultural mega-events in order to garner the fleeting attention of the world's media is indicative of the ways in which the economic survival of cities is implicated within ever more integrated circuits of finance capital, corporate marketing, multi-level governance and international relations.

Prior to the end of the Cold War it was still possible to point to city–regions in 'actually existing socialist states' where the circuits of economic power were rather different, and where the central organisation of political and bureaucratic power continued to shape and determine the production and consumption patterns of urban workers and citizens. Today, a very few such cities exist even in economies such as China's which have retained a strongly authoritarian political system.

However, as we noted in the previous chapter, even in the most pro-market democracies such as the United States, we can see that the pattern is far from even, that the existence of business interests and urban regimes is not always consistent with a pro-growth agenda, that enterprise is almost invariably dependent on governments smoothing its way in the form of grants, social wages, sympathetic planning decisions and tax breaks. This all gives credence to David Harvey's observation, following Robert Goodman (1979), that state and local governments are the 'last entrepreneurs' (Harvey 1989a: 126) – a status that has been reinforced by the unprecedented scale of financial bailouts paid to banks, insurance and mortgage companies and the auto industry during the global fiscal crisis of 2007–9.

As cities and regions begin to experience the seismic impact of the global financial crisis, there has been a noticeable shift in urban policy circles towards a much more critical view of neo-liberal anti-statism and a growing appreciation among urban workforces that the best cities in which to live in the middle of a financial downturn are those where public sector employment levels are high, and collective public goods – such as schools, hospitals and social services – are still maintained to a good standard. What is not in doubt, however, is that geographical mobility, whether voluntary or not, is increasingly defining the mosaic of urban society, and giving rise to a bewildering variety of communities and social identities – which nevertheless succeed in maintaining some consistency and valency in an increasingly transnational and hybrid urban world.

In the three chapters that comprise the following part of this volume we consider the extent to which personal and collective identities shape and are shaped by the governmentalities of cities, how and why the management and dissemination of information and communication is central to any debate on the nature of urban power, and the ways in which the production and use of urban space and the built environment order and regulate the lives of urbanites in profound and significant ways.

Further reading

Social Theory and the Urban Question by Peter Saunders continues to be an influential and important reference for students and scholars alike, despite its controversial and somewhat iconoclastic approach to the prevailing orthodoxies of the time – especially in relation to the idea of a 'non-spatial urban sociology' (Saunders 1986). See also his 'Space, the City and Social Theory' in Gregory and Urry's *Social Relations and Spatial Structures* (Gregory and Urry 1985).

A key text for any student interested in the early debates on the spatial characteristics of Fordism and post-Fordism remains Ash Amin's edited volume *Post-Fordism: A reader* (Amin 1990). The articles by Deas and Ward (2000) and MacLeod (2001) are especially useful in providing a European focus on 'new regionalism', while Brenner provides a telling critique of the US literature (Brenner 2002).

The literature on the 'scalar turn' in political and economic geography, regional studies and sociology has grown significantly in recent years and is moving from a predominantly European–North American focus to the 'new state spaces' of the Asia–Pacific region, Latin America, the former Soviet Union and eastern Europe. An excellent discussion of the scalar turn can be found in Keil and Mahon's *Leviathan Undone? Towards a political economy of scale* (Keil and Mahon 2009).

The defining statement of the creative cities thesis is to be found in Richard Florida's 2003 paper 'Cities and the Creative Class', which formed the core principles of his larger study (Florida 2005). For a critical rejoinder, Jamie Peck's article (Peck 2005), 'Struggling with the Creative Class' is widely cited by sceptics of the creative-cities literature.

Part IV
Identity, communication and space

7 The politics of urban identity

> beneath the fabricating and universal writing of technology, opaque and stubborn places remain. The revolutions of history, economic mutations, demographic mixtures lie in layers within it, and remain there, hidden in customs, rites, and spatial practices.
>
> Michel de Certeau, *The Practice of Everyday Life*

Introduction

Despite the predictions of many twentieth-century social theorists that the age of modernity would lead to the emergence of an undifferentiated mass society in which the consumer economy would entirely replace antecedent value systems based around class, religion, culture and ethnicity, these collective identifiers – or what the late cultural theorist Raymond Williams called 'structures of feeling' – have proved remarkably enduring (Williams 1961). Indeed, far from ushering in a new age of global liberalism, as Francis Fukyama predicted, 'the end of history' has witnessed a dramatic growth in identity-based groups and movements – especially in cities and towns around the world that have recent experience of conflict or where large-scale migration has occurred, or where the population is becoming more socially and culturally diverse.

The reasons for this trend are numerous and complex, but among the most important, one would have to point to the establishment of major colonial empires and colonial metropolises in the eighteenth and nineteenth centuries. This territorial expansion was accompanied by a vast traffic in human slaves and indentured labour both within and outwith the colonial economy. Major intercontinental population migrations to the Americas in the latter nineteenth and early twentieth centuries were paralleled in the same period by large-scale intra-regional migrations in Europe, Asia, the Far East, Arabia, northern and sub-Saharan Africa and Russia and its bordering territories. Two world wars and the economic catastrophe of the Great Depression of the 1930s induced still more migration from those fleeing violence and persecution to those in search of a livelihood and an alternative to starvation. In the aftermath of the Second World War, the rapid growth in independence movements and self-rule in the former European colonies did not diminish the political, cultural and economic ties to 'the motherland' whose cities became a more attractive destination for the children of empire as the post-war boom offered the prospects of jobs, education and a new life. With the collapse of 'actually existing socialism' in the Soviet Union and eastern Europe at the end of the 1980s and early 1990s the polarities of the Cold War that had oriented the nations of the world either to 'the West', 'the East' or a non-aligned position were suddenly switched off.

In a monopolar world dominated by the United States whose major economic corporations have continued to strengthen their pre-eminence and control of global markets since the 1980s and whose military complex has extended and consolidated its presence in every region of the world, for anti-capitalist and socialist-type societies the 'end of history' appeared to be a fitting epithet. However, western economic and military dominance has not brought with it an acceptance and enthusiasm for the social, cultural and moral values on which an increasingly neo-liberal state-sponsored capitalism relies in many parts of the world. This is especially true for the cities and urban settlements of the Global South, which Mike Davis has referred to as 'the planet of slums', and where up to 1 billion people now live (Davis 2006: 26).

In other words, modernity has not led to the homogenisation of urban populations around the world but to an ever more intensive social differentiation and segregation (Appadurai 1996, 2001). The vast majority who are subject to this global spatial sorting are entirely powerless to control it – their social identity is decided by the arbitrary circumstance of having been born to the wrong mother, in the wrong place, at the wrong time (Williams 1981). Thus when we talk of social identity, whether in an urban setting or not, for most of the world's population their identity is one that is *negatively ascribed*, and there is little or no possibility of altering it. If we bear in mind that four-fifths of the world's population live in poverty, that the majority of this population are female, and that women experience greater degrees of poverty than men, then it is quickly obvious that while poor women do make up the majority of the world's urban population, their collective identity is not a source of power but a badge of social inequality.

For the remaining affluent fifth that mostly live in western towns and cities, social identity is generally an *elective condition* that comprises socio-economic status, ethnicity, gender, age, beliefs and values, education, consumption patterns and, increasingly, residential location. Of course there are, even in developed urban societies, many people for whom – by virtue of their socio-economic status – the degree of choice over where they live, what products they are able to consume, and even what values and beliefs to hold will be tightly constrained. The internalisation of negatively ascribed identities can be as limiting of the rights to equal participation in the life of the city for marginalised populations in the West as it is in the Global South.

Having dealt with the emergence of collective political actors in Part I of this volume as essentially instrumental or goal-oriented movements, in this chapter we explore how different political processes and structures help to shape, to maintain and also to contest forms of urban identity in different urban and regional contexts. No social milieu that leads to what sociologists term 'norms and values' emerges spontaneously and without some relationship to the institutional framework within which the reproduction of social life takes place. However, disentangling 'the hidden wiring' of modern urban identity can be a daunting task, and what follows reflects some of the new research that is attempting at least to map the distribution boards of these emergent forms of the socio-spatial habitus.

Identity, difference and resistance

> I've come more and more to prefer the term 'identification' to the term 'identity'. Only because identity suggests something fairly formed, fairly fixed, fairly exclusive, fairly stable. In the classical sense of the world people have many identities . . . But if you move to identification, you move to a kind of process, where people are engaging this menu of possibilities in the work of the imagination. In and through the work of the imagination, they are, as it were, trying out many possibilities – and in many cases they're forced to try out some possibilities. I'm very conscious . . . of the brutal ethnic violence in Gujarat, where there's a continuing state-condoned and even state-sponsored pogrom which began in late February 2002 and is worse than anything since the partition of India in 1947. So I'm keenly aware that the identity that goes under the label 'Hindu' or 'Muslim' is not something most people can explore – they have no choice here.
>
> Appadurai 2002: 43–4

Appadurai makes a valid point when he associates identity with a static individual or collective ontology, but in the context of Gujarat it is clear that a particular religious identity is not only imagined but is rather a non-negotiable marker of belonging and difference. Therefore to the extent that social identities are largely predetermined by one's membership of a self-similar group or community we could say that identity proceeds from difference. Due to their much greater concentrations of population than rural or semi-rural habitats, cities are more likely to contain a greater degree of social, ethnic, religious and cultural diversity. But the existence of a greater heterogeneity does not mean that social equality is necessarily more in evidence in cities. In fact, those who share a collective identity that is negatively regarded by mainstream society often find that their exclusion from social, economic and political life is more intensified in an urban environment that in western and colonial societies is designed by and for the holders of economic and political power.

Brendan Gleeson talks about how the oppression of disabled people 'takes a distinctive form in cities and that certain general urban characteristics – notably urban design, employment patterns, and the distribution of land uses – entrench social discrimination against disabled people' (Gleeson 1998: 91). Neither does membership of an 'othered' group offer the prospect for a greater degree of tolerance and respect to those who are also 'other' to the dominant identities and values of the sub-group. When the figure of the 'black dandy' emerged in Harlem in the 1920s and 1930s the contested identities of race, sexuality and masculinity came together in the form of an 'openly queer desire and anti-bourgeois politics' that at once challenged popular notions of what it means to be black in New York during the depression years, but also the dominant value systems (both black and white) that regard any form of sexual or racial hybridism as decadent and morally corrupting (Glick 2003: 414). Equally, since the events of 11 September 2001, forms of Muslim religious identity that involve covering have become a contested ground in which women's bodies have been claimed as a symbolic site for the assertion of moral and religious values that are consciously 'anti-western' and by those civil society and state actors who regard veiling and covering as socially and politically divisive and a challenge to universal civic values including gender equality (Werbner 2004; Pak 2006; Jusova 2008).

The city of Antwerp provides a useful example of how identity and culture can provide both a vehicle for collective self-expression and a site of contestation. This Belgian city has been a stronghold of the Flemish nationalist movement, Vlaams Belang (VB, meaning Flemish Interest), for a number of years, but it is also home to a large, predominantly North African population. In the 2006 municipal elections, VB retained around one-third of the vote, but was excluded from power by a coalition of centre-left parties. VB's supporters see themselves as Flemish nationalists who are intent on defending their language and culture against both the francophone population of Belgium and what they consider to be the 'Islamisation' of cities such as Antwerp by Belgium's Muslim communities. Each time a new mosque is proposed in the city – 36 currently exist – Flemish Interest supporters organise a counter-protest. The VB has also used

Antwerp as a rallying ground for far-right groups from all over Europe whose declared aims are to resist 'creeping Islamisation'.

An example of what such groups object to was the local council's decision to ensure that all food in city schools would be halal. However, previously the local authority had alienated many of Antwerp's Muslim population by issuing a ban on city employees wearing the hijab when dealing with members of the public – although nursery workers were allowed to wear a bandana as an alternative and staff not in contact with the public were permitted to wear a headscarf.[35] The local authority defended its decision on the grounds that members of the public had the right to be treated neutrally and that the display of 'visible symbols of philosophical, religious, political or other opinions' might alienate some city residents.

As Coene and Longman write, 'On the political and judicial level, the principle of gender equality and women's rights seems to point in opposite directions, with hijab bans being defended in terms of the right of protection against gendered oppression, and protests against these bans emphasizing Muslim women's rights to religious freedom and personal choice' (Coene and Longman 2008: 303). The Belgian case differs from that of France where an attempt to ban the wearing of religious regalia in 1994 served to reinforce the uncompromising secularism that has long been a central tenet of the republican state (Vivian 1999). In Belgium the much weaker position of the federal state and the re-territorialisation of a number of important policy domains in the wake of demands for greater autonomy by both the Flemish and the Walloon populations has strengthened the powers of city leaders in defining what the rights and obligations of citizens should be in an increasingly multi-ethnic society. In other words, the 'menu of possibilities' that determines how contested collective identities are imagined, defined and negotiated tends to be written by the restaurateur rather than by those who come to dinner.

Reflexivity, modernity and urban identity

If social identity from the Middle Ages to the industrial revolution was defined by the phrase 'you are what you do', then its equivalent in the digital age must surely be 'you are your network'. Indeed so rapidly has this insight been assimilated into popular culture that mobile phone companies base their entire advertising campaigns on the assumption that we are not only comfortable with the notion of our ontology consisting of a constant flow of data between friends, family and work colleagues – but that we positively welcome this and want our networked selves to grow and to be 'always on' or constantly available.[36]

The possibility for self-reflexivity – the democratisation of the 'mirror of princes' that we discussed in Chapter 2 – was only made possible as a result of an urban sensibility that required an ability to self-regulate one's conduct in public and private. Initially a courtly virtue, the expression of good taste and a refined knowledge of literature, philosophy and the *beaux arts*, and the ability to talk convincingly about new developments in architecture, design and fashion were performative requirements for any member (or aspiring member) of high society. The dissemination of these courtly virtues, as Richard Sennett (1990) details in *The Conscience of the Eye*, to the urban bourgeoisie in the seventeenth and eighteenth centuries can be traced directly to the phenomenon of the *flâneur*, who is such a central feature of Baudelaire's poetry and whom Walter Benjamin saw as the harbinger of modernity in nineteenth-century Paris (Benjamin 1973).

The Swedish anthropologist Ulf Hannerz, in his influential book *Exploring the City*, demonstrates the ways in which community and identity can be influential on urban structures (Blokland and Savage 2001: 222). Drawing on many classic works of twentieth-century urban anthropology his basic premise is that society (and in particular urban society) can be studied through a series of situations that he describes as a *role inventory* or the total array of modes of behaviour within a major social unit – such as an urban community. Individuals are involved in a series of modes of behaviour that constitute their *role repertoire* and they perform these different roles within discrete domains (the household, the neighbourhood, etc.). How people are 'recruited into situations' depends on what he describes using the clumsy term of *role-discriminatory attributes*, or 'culturally defined characteristics of individuals which exist apart from particular situations' such as sex, age, ethnicity or race (Hannerz 1980: 317).

Hannerz provides an example of how such role-discriminatory attributes – or what I have preferred to describe elsewhere as collective identities (Parker 2001b) – can optimise access to social power in western African towns where the shared background of the members of a voluntary association of migrants 'provides not only a sense of trust but also one of solidarity' (Hannerz 1980: 156). These types of associations are not found everywhere in urban Africa – or indeed in more developed societies among incomer populations – but they have not entirely disappeared either.

The work of Hannerz, along with that of earlier social theorists including Simmel (1950) and more contemporary analysts such as White (2008), questions the idea which had been firmly established by the Chicago school of urban sociology through the concept of the 'folk–urban continuum' (Miner 1952) that strong collective beliefs and identities are exclusively the preserve of rural communities and instead shows why identity-based social networks have continued to be central to the life of cities.

The politics of the urban habitus: the elective affinities of place

Several decades ago the sociologist Richard Sennett wrote a book whose subtitle 'Personal Identity and City Life' was a rather better clue to the subject matter than its more enigmatic title, 'The Uses of Disorder'. Sennett sought to explore 'the vague term of "community"' in terms of its contemporary relevance to highly urbanised and affluent population groups:

> What does it mean for a white, educated, affluent person to feel a sense of community with other people? . . . the first generation that has lived with both the achievement of affluence as a constant force in life, and the problems of what to do with it . . . has no model from the generation that brought the affluence into being, since the wilful innocence of the suburbs does not seem to be a satisfying way to sustain a social life.
>
> (Sennett 1970: xi–xii)

Building on the work of David Riesman and the early Chicago School pioneer Florian Znaniecki, Sennett goes on to argue that there is something adolescent in the desire of newly affluent suburbanites to pretend to a common set of experiences that they share in order to create an imagined community 'that binds them all together as one being, with a definite set of desires, dislikes and goals' – in this way '[t]he image of the community is purified of all that might convey a feeling of difference, let alone conflict, in who "we" are' (Sennett 1970: 36).

Mike Savage and his colleagues have suggested that central to what has been termed 'the spatialisation of class', the idea that occupation and employment position *in and of themselves* continue to be the fundamental units of class analysis needs urgent reappraisal (Savage et al. 2004; Parker et al. 2007). This is because social stratification is increasingly seen as a fundamentally spatial process. For example, Savage et al. conclude:

> One's residence is a crucial, possibly the crucial identifier of who you are. The sorting processes by which people chose to live in certain places and others leave is at the heart of contemporary battles over social distinction. Rather than seeing wider social identities as arising out of the field of employment it would be more promising to examine their relationship to residential location.
>
> (Savage et al. 2004: 207)

In other words, the suggestion here is that where we live is more socially and culturally salient than our employment status. Pioneering urban sociologists such as Ray Pahl should be credited with identifying the collective locational choices of middle-class residents – what he termed 'where consciousness' – as a key issue for urban social theory 40 years previously (Pahl 1970: 219). More recently it has been the field of 'commercial sociology' (Burrows and Gane 2006) and 'policy sociologists' working on neighbourhood-based classificatory schemas premised upon the empirical observation that 'people tend to live with others like themselves, sharing similar demographics, lifestyles and values' (Weiss 2000: 305) that has led to the development of new methodologies and the generation of rich empirical data in this area (Savage and Burrows 2007; Savage and Burrows 2009).

The spatially patterned clustering of self-similar socio-types at the urban and regional level has given rise to an important and highly lucrative industry. Claritas (which owns PRIZM, one of the most widely used commercial neighbourhood classification systems in the USA) claims that its products are based upon 'the fundamental sociological truism that "birds of a feather flock together" . . . [and] . . . "You are where you live"' (quoted in Goss 1995: 134). This proposition would appear to have antecedents in the Chicago school of urban ecology (Park et al. 1925) and the urban sociology of Rex et al. (1967), Pahl (1970) and others on 'housing classes'. But in many respects the idea of identity-based spatialities – or 'class places' – derives from applications of the work by Pierre Bourdieu on 'capital', 'habitus' and 'field' as a means of interpreting the preferences, tastes, strategies and actions of the urban bourgeoisie (Bourdieu 1986, 2000, 2005).

According to Savage et al., spatial clustering occurs among people that enjoy mobility choices because they, 'are comfortable when there is a correspondence between habitus and field . . . otherwise people feel ill at ease and seek to move – socially and spatially – so that their discomfort is relieved . . . mobility is driven as people, with their relatively fixed habitus, both move between fields . . . and move to places within fields where they feel more comfortable' (Savage et al. 2004: 9). With the advent of interactive Internet-based information services (known generically as Web 2.0), highly geographically mobile and affluent populations now have a wealth of neighbourhood-specific quantitative and qualitative data at their fingertips, which they can sort according to their own personal or household identity matrix (including education level, religious faith, income bracket, cultural and leisure interests, political beliefs, ethnicity, age and so forth).

In North America, websites such as Bruce Sperling's 'Best Places' encourage re-locaters to make use of urban performance indicators in terms of crime, high-school graduation rates, cost of living and even the prevalence of singles bars in order to select the residential habitat that most closely matches their consumer preferences. As the leading British geodemographer, and architect of the Mosaic social classification system, Richard Webber comments '[t]hat such terms [category labels and descriptors] were coined without any awareness of or exposure to the academic debate on globalisation and gentrification suggests that the processes and patterns identified qualitatively by urban sociologists map very closely with the patterns picked up quantitatively by the data mining techniques of marketers' (Webber 2007). In other words, those working within the commercial sector on ever more accurate and meaningful market segment classifications at the neighbourhood level are revealing the extent to which cultural self-reflexivity and spatial location choice has become highly structurated and automated on a global scale (Thrift and French 2002).

Scott Lash views contemporary social geography as becoming increasingly fragmented as a result of the interplay between two main drivers – the variable density of 'information flows' and the prior nature of the 'identity spaces' that such flows envelop. Lash draws a distinction between what he calls 'live' zones and 'dead' zones in the fluid infoscapes that are emerging across the globe. Live zones are where such flows are at their most dense and dead zones are where the flows are lightest.

However, for Lash, this contemporary infoscape intersects in variable ways with zones of another sort – what he terms 'tame' zones and 'wild' zones. He writes:

> Live zones and dead zones refer most of all to economic spaces (though of course to an increasingly 'semiotic' economy), whereas tame zones and wild zones refer most of all to 'identity spaces' . . . the live and dead zones of economic spaces refer to the presence (or relative absence) of the flows and the identity spaces refer to what social actors do with them.
>
> (Lash 2002: 28–9)

The idea of social identity being no longer an ascribed condition but a spatially articulated and articulate property of knowledge bearing agents is consistent with the medieval notion of 'Stadt Luft macht frei' – city air makes you free – since Lash's wild and live zones are unlikely to be found outside the metropolis. However, the idea that cities are necessarily harbingers of cosmopolitan, fluid and elective identities is very much a recent feature of certain western urban societies. As we have previously noted, in the rest of the world and throughout most periods of history, one's idea of selfhood was rarely a matter of choice so much as – in the case of religious and ethnic minorities – a question of survival and preserving a particular form of elective belonging in the face of often very hostile values and cultural norms.

Little attention, however, appears to have been given to the urban context in which the politicisation of cultural identities takes place. Rather it is often assumed that the coexistence of diverse ethnic and cultural identities in cities is simply the result of transnational migrational flows and that the attraction of western cities to new migrants lies principally in the economic opportunities that derive from the high concentration of jobs and services. Much less attention has been given to the idea that cosmopolitan cities are increasingly places of identity *affirmation* rather than the identity *erasure* of fin-de-siècle Berlin associated with Simmel's anonymous crowd and the 'blasé attitude' (Simmel 1950).

Identity, conflict and place

A common experience or perception of negative difference allied to the concentration and segregation of discrete population groups according to ethnicity, socio-economic status and housing tenure creates the potential for the negative assertion of identity through violence and conflict as a form of collective 'voice' when feelings of 'loyalty' have been betrayed by discrimination and 'exit' is a route open to only the most wealthy and 'non visible minorities' (Hirschman 1972).

In the summer of 2001, several major towns in northern England[37] became the focus of violent street protests, which the reports into the disturbances largely attributed to racial segregation and a lack of 'social cohesion' (Worley 2005: 484–5). A major report into the disturbances in Oldham was published four years after the events with the aim of investigating how far attitudes and behaviour had changed among the town's Asian and white communities. The report noted:

> In a borough where neighbourhood and district identity and allegiance commonly take precedence over identification with the borough as a whole, it is not difficult to see how this could act as a major constraint in building community cohesion and tackling segregation.
>
> (Cantle 2001: 47)

The review team's overall impression 'was that segregation and divisions between Oldham communities is still deeply entrenched'. The report went on to comment:

> This is as much in the minds of people as in neighbourhood structures and is at odds with experience in many other areas of the country. Hence our view that if you want to change a community, the community must want to change. In short, polarised communities continue to be a significant feature of relations across all sections of Oldham society. For example, a young Muslim mother told us:
>
> > 'My neighbour is Indian and my Muslim community tell me off for speaking with her. They say, I should speak to her if I am getting her to embrace Islam otherwise no.'
>
> Similarly, a young white male told us that:

> 'I have nothing to do with them (Asians) at my college. We have nothing in common and we would not want to get involved with each other. We *are happy as we are*'.
>
> (Cantle 2001: 49)

In Northern Ireland, the advent of the paramilitary ceasefire in 1994 did not lead to the cessation of intra-communal hostility. According to a report compiled by the Institute for Conflict Research, 'Northern Ireland Office figures indicate that interface barriers remain a presence in many urban areas and that at least 17 barriers have been built, extended or heightened in Belfast since the ceasefires of 1994.' Indeed in a more recent report the same author notes the persistence of no fewer than 87 security barriers (including those constructed by the police service), which have been erected since the 1970s in interface areas where Catholic and Protestant housing are closely sited to each other. Only five barriers have removed since the peace process, and many of these only as a result of commercial redevelopment (Jarman 2008).

Limited power sharing between republicans and unionists at a province-wide level has not been matched by local community coalitions on the ground. The report's author notes 'the limited focus given to the subject of sectarian violence by civil society organisations' (Jarman 2005: 4) in Northern Ireland's urban communities. As with English cities such as Oldham,

> While many [local community organisations] would claim to be anti-sectarian, much of such work takes place within a single identity context. Any substantial cross-community activity against sectarian violence only seems to occur in response to specific and horrific acts, but such reactions have rarely been sustained. It would be beneficial if some of the umbrella civil society organisations developed a more sustained campaign around this issue, as has begun to develop in response to racist violence.
>
> (Jarman 2005: 6)

The North Belfast Interface Network (NBIN) is a rare example of just such a community organisation. Founded in 2002 by local community organisations from both Catholic and Protestant neighbourhoods in north Belfast, which had experienced high levels of sectarian violence in the past, the NBIN's aim was to 'develop a strategic response to interface violence and develop community relations work in North Belfast'. As one of the group's contributors, Breandán Clarke, wrote on the tenth anniversary of the Good Friday Agreement, 'Walls themselves are neutral, inanimate structures, they hold no politics, intent or opinion but the use of a wall to separate, to divide, to partition and to control is laden with the politics of its architect'.[38]

Clarke agrees that 'the effective solution to the removal of the walls lies within the communities that are still kept apart from each other by their presence', but

Figure 12. A Protestant Orange Order Lodge marches on 12 July to commemorate the victory of King William of Orange during the Battle of the Boyne in 1690. © Discodave2000/Dreamstime.com

he also notes that in the 10 years since the Good Friday Agreement 'no plan has been put forward by government to address the interface walls or the associated sectarianism they foster'.[39] Both nationalist and unionist politicians have a vested interest in maintaining and strengthening strong ties between religious confession, political allegiance and spatial location (see Figure 12). Although both Stormont and city governments pay lip-service to the importance of 'community integration', there has been little attempt to promote integrated education through non-denominational schooling or public housing developments.

Conclusion

Urban populations with strong but highly sectarian identities such as those in Northern Ireland or Mostar or Jerusalem are not advantaged by the 'strength of weak ties' (Granovetter 1973) that Richard Florida identifies in hip, happening cities such as Austin or Toronto, but by what (inverting Granovetter's thesis) we might call the 'weakness of strong ties' involving a defensive essentialisation of collective identity markers, rituals, customs and spatial dispositions that generates hostility and antagonism towards non-in group communities and a reluctance to identify with the metropolitan as a scale of belonging.

As we noted in Chapters 3 and 4, political identities can exist as latent cleavages within urban society – as markers of class, religious belief, gender, race and ethnicity – but they only assume the characteristics of power assemblages once they have been articulated, mobilised and strategically deployed by political hegemons. At the same time, the state plays an important role in affirming or denying the value of discrete collective identities, and establishes modes of governmentality by which identity is encoded and sorted according to the policy algorithms which allow or deny populations access to public resources, social status and economic well-being. This social coding and sorting relies on information and communication technologies that both represent and construct the urban worlds – the mediated city – from which our sense of belonging and otherness derives. In the following chapter we explore why information and communication need to be understood in terms of both the ontology of cities and as epistemic routines that determine how cities operate and relate with other spatialisations and articulations of power/knowledge in the interstices between the global and local, and the citizen and the state.

Further reading

Arjun Appadurai's book *Transurbanism* (Appadurai 2002) is an evocative account of the emergence of a global urban citizenry that is both strongly attached to cultural and indigenous identities and practices while at the same time capable of negotiating a variety of urban contexts that offer different challenges and opportunities to an increasingly mobile population.

Globalisation and Belonging by Savage, Bagnall and Longhurst (Savage et al. 2004), is focused mainly on the UK, but its argument applies equally well to many of the other national urban/regional locales discussed in this chapter. The book's insight is particularly important because it signals a move away from occupational-based analyses of social identity in urban sociology to notions of belonging relating to residential location and consumption. In a similar vein, Paul Kennedy's *Local Lives and Global Transformations* (Kennedy 2010) deals with aspects of urban and regional identity in the context of globalisation using well illustrated local case studies drawn from around the world to aid the discussion. Scott Lash's *Critique of Information* (Lash 2002) is an influential and pioneering study of the relationship between socio-spatial identities and information-based capitalism in a global urbanised economy.

8 Information, communication and the networks of urban power

> The use of any kind of medium or extension of man alters the patterns of interdependence among people, as it alters the ratios of our senses.
>
> Marshall McLuhan, *Understanding Media*

> The city is a medium.
>
> F. Kittler and M. Griffin 1996

Introduction

Communications, be they physical (roads, waterways, railways, air routes) or media-based (post, telegraphs, telephones, terrestrial and satellite broadcasts, and data networks), have been responsible for the emergence of cities as places of human settlement and for their key role as producers and relayers of information both for local and increasingly regional, national and global consumption (McLuhan 2001: 97). The strategic importance of communication during times of war is well known – and this is why cities are frequently targeted for aerial bombardment. Indeed during the NATO bombing of Serbia the coalition forces deliberately used 'surgical strikes' against the country's television and radio infrastructure as well as military radar and communication facilities in an effort to remove the government's ability to communicate with the population.[40]

Sheller and Urry rightly point to the tendency of urban studies to rely on a 'place based ethnography', which has 'in some ways limited the horizons of urban studies to what can be seen "on the streets"' – a critique that we shall also explore in the following chapter. Rather the authors argue, we need to ask, '[h]ow are cities being de-materialized and re-materialized through new kinds of urban logics, technical systems and discursive orderings'(Sheller and Urry 2006: 4). But at the same time these new networked spaces are not simply passive receptors of communication and information signals; they also induce and produce physical mobilities in time and space that in turn generate the need for material, physical communication infrastructures (airport expansions, satellite relay stations, data archiving and processing centres and so forth) (Graham and Marvin 2001; Sheller and Urry, 2006).

When thinking about information and communication technology *and* just as importantly the content of the media that is produced at an extraordinary rate in (if not always exclusively for) the world's cities it is important to think about the ways in which power has historically been asserted, sustained or resisted through the use of information and communications media and how emergent technologies (especially digital technologies) are providing a much greater potential for surveillance, personal and aggregate data gathering and analysis, data mapping, sorting and discrimination.

We need therefore to consider the question of information and communication within and between urban networks as both a 'power over', for example, the ownership of media corporations and hence (to a large extent) the content of information and communication flows; and, second, in terms of 'power to', for example, the power to control access to public or private services based on one's residential location. There is also a third aspect of power that this chapter will consider, which is related to some of our earlier discussion on governmentality, and which is concerned with what has been broadly described as 'biopolitics'. Here we are interested in the state's capacity (both the local state and the national state acting locally) to record, store and monitor information on citizens and visitors to the city and to *surveille* communications between subjects and groups and organisations for a whole range of explicit and non-explicit purposes.

The city as medium

> The city and the home in the tribal world (as in China and India today) can be accepted as iconic embodiments of the word, the divine mythos, the universal aspiration. Even in our present electric age, many people yearn for this inclusive strategy of acquiring significance for their own private and isolated beings.
>
> McLuhan 2001: 134

Authors such as McLuhan, Kittler and Mumford have drawn attention to the fact that the city has long existed as a *medium*. In the period after the Second World War, as Scott McQuire observes, 'the interlacing of urban space with high-speed interactive networks . . . constitutes a critical change in urban experience'. This critical change is manifest in terms of the 'dispersal of economic activities across geographical space, increasingly on a global scale', as well as the production of, 'increasing concentrations of power, as command and control centres for the global economy are consolidated in relatively few "global cities"'(McQuire 2008: 20–1). McQuire wants to argue that terms such as 'the informational city' or 'the digital city' fail to encompass as well as his preferred term of 'the media city', 'the historical dimensions of the relation between media and modern urban space'. As information and communication technology becomes 'increasingly mobile, scalable and interactive' the media city becomes defined in terms of what McQuire calls 'relational space'. Drawing on Scott Lash's notion of the shift from closed personal bodies to open social bodies and Ulrich Beck's concept of 'reflexive modernity', the relational space of the media city provides the manifestation of this 'ambivalence of mobile spatial configurations and ephemeral individual choices' (McQuire 2008: 22).

McQuire is essentially arguing that the variable speeds and intensities of distanciated forms of communication and information are radically reconditioning and re-inscribing the urban experience such that '[t]his heightened volatility has increasingly become an operative factor in the exercise of power'. The 'rhizomic' nature of contemporary networked information distribution and exchange (Deleuze 1976; Deleuze and Guattari 1983b) induces power to spring up in 'less obvious tangents' such that, in Kittler's terms,

> Power means occupying at the right moment the channels for technological data processing. And centrality becomes a variable dependent on media functions, rather than vice versa.
>
> (Kittler and Griffin 1996 in McQuire 2008: 23–4)

But while this may be true in terms of the location of NASDAQ behemoths like Google in new media regional economies such as Silicon Valley, it is important not to lose sight of the fact that much of the world's media is produced and consumed locally – even though the control of local urban media operations is increasingly in the hands of a small group of globally dominant media and communication giants that are still very much concerned with the cultivation and infiltration of traditional sites of political power.

The fact that at its peak in 1999, 'over half of all venture capital in the USA poured into the communications and media sector' (Picard 2002: 175 in Winseck 2008: 40) is testimony to the importance of information and communications not only as a lucrative investment opportunity but also as a key site of ideological power that dates from the earliest investments in local news sheets to the Fordist domination of media ownership such as *La Stampa* in Turin (home of the Agnelli family FIAT dynasty) to post-Fordist 'new money/new media' ventures such as Berlusconi's Mediaset empire based in Milan – which initially began as a local cable TV network venture before becoming a national and then international multi-media conglomerate (Ginsborg 2005).

Cities and the ownership of the means of communication

As Arsenault and Castells observe:

> Today, media are organized around a global network of multi-media corporations that extend from a core of diversified multi-national media organizations, to large national and regional companies, and to their local affiliates in different areas of the world.
>
> (Arsenault and Castells 2008: 707)

Cities that are globally significant in relation to service concentration also tend to be significant in terms of their contribution of globally important media operations, although the global media are less dispersed than the global service industries in the major world cities (Krätke and Taylor 2004). Interestingly, two of the 'big three' global cities – London and New York – led the pack in the 2000s in terms of both service and media concentration respectively (see Figure 13).

In the United States the seven major media conglomerates include Disney, CBS, AOL-Time Warner, News Corp, Bertelsmann AG, Viacom and General Electric, and together they own more than 90 per cent of the media market. Most of this market is concentrated in towns and cities, where despite a proliferation of local TV and radio stations there are very few genuinely local and independent owners left – a scenario that applies equally to the press. According to the US Federal Communications Commission and the Telecommunications Act of 1996, a single company can now own up to 45 per cent of any given market whereas in 1927 the

Figure 13. The top 10 connected business service cities in terms of media connectivity by ranks. Adapted from Krätke and Taylor 2004.

Service rank	City	Media rank	Difference
1	London	1	0
2	New York	2	0
3	Hong Kong	20	–17
4	Paris	3	+1
5	Tokyo	18	–13
6	Singapore	12	–6
7	Chicago	24	–17
8	Milan	5	+3
9	Los Angeles	4	+5
10	Toronto	8	+2

limit was set at 25 per cent.[41] Furthermore, 98 per cent of American cities have only one daily newspaper and a growing number of even quite large towns now have none at all, which research has shown leads to less political participation and democratic scrutiny (Baker 2002: 28).[42]

In the view of one critic, 'the most common assumption is that the owners of the media influence the content and form of media content through their decisions to employ certain personnel, by funding special projects, and by providing a media platform for ideological interest groups'. Quoting the work of British media researchers Curran and Seaton (2003) Meier reports that 'the national press generally endorses the basic tenets of the capitalist system – private enterprise, profit, the free market and the rights of property ownership'. Because of the increasing vertical integration of media outlets, what holds true for the 'national press' also applies to the urban and regional media, but because of the global concentration of media enterprises the notion of a national, regional or local press/media is meaningful only in terms of readership. Meier goes on to observe:

> In the United States . . . a small group of powerful owners of six to ten media conglomerates, control what is read by the population, what people see and hear – or do not read, see and hear. Concerns are expressed about increasing corporate control of mass mediated information flows and about how democracy can function if the information that citizens rely on is tainted by the influence of mega-media.
>
> (Meier 2002: 300)

Such concerns are not only limited to media ownership patterns in the cities of the Global North. According to Nyanmnjoh, 'the media in Africa are effectively controlled by government and capital, who are both keen to feed the public with nothing subversive to their interests and power' (Nyanmnjoh 2004: 128). He identifies the twin perils to media freedom of centralised government control – which extends to all the major African cities with few exceptions – and the increasing consolidation of the African media market as structural adjustment programmes favour highly capitalised oligopolistic media conglomerates that are prepared to trade market dominance for political interference with editorial content. Meanwhile, in large parts of China,

> the shift to business subsidy through advertisement to media firms has been mainly responsible for the relative decline of the traditional monolithic national and provincial Party organs, and the rise of urban mass appeal newspapers . . . and broadcast channels. The same process has also been largely responsible for . . . the marginalization of media outlets catering to social groups who are in the majority in numbers, but marginal in their political and economic power: workers, farmers, and poor women.
>
> (Zhao 2004: 187–8)

The internationalisation/globalisation of both media networks and content has seen 'local and regional players . . . actively importing and/or re-appropriating foreign products and formats while corporate transnational media organizations are pursuing local partners to deliver customized content to audiences' (Arsenault and Castells 2008: 708). Bennett reports evidence of similar trends 'emerging in many nations as local media are purchased by global media giants . . . Since few models of democracy advocate communicating with citizens according to their commercial viability', he argues, 'this might seem to be an inherent problem with deregulated commercial news media' (Bennett 2004: 141).

A similar pattern of corporate concentration can be seen in relation to electronic media usage and especially Internet usage, where in the United States, after Google, Microsoft and Yahoo (the three biggest search engine providers), AOL-Time Warner, News Corp (which owns MySpace) and Disney are all positioned inside the top 10 parent companies for consumer reach and monthly online time by American households.[43] Affluent cities on the west and east coast of the United States continued to lead the way in terms of American Internet use during the first decade of the twenty-first century. According to a Reuters report,

> [t]he cities that rank highly for broadband penetration are also prominent Internet usage markets. For example, adults in San Francisco, Boston and San Diego are more likely than the average person to have accessed the Internet during the past month, and they are also more likely to have spent 10 or more hours online during the past week. San Francisco adults are 12 percent more likely than all adults nationally to have accessed the Internet in the past month, and 26 percent more likely to have spent 10 or more hours online during the past week.[44]

But, despite the relative proliferation of broadband take-up in a small number of 'silicon cities' in the United States, OECD data revealed that the USA ranked only 15 out of 30 OECD countries surveyed in 2008 and that its position had been falling relative to the rest of the world since 2001. Despite former Vice-President Al Gore's promise that the American economy would be reborn along the information superhighway, the biggest obstacle to US dominance as an urban information-based economy turns out to have been a combination of weak government support for strategic information infrastructure (such as fibre-optic cable networks) and the increasing flight of the American population to suburban and increasingly ex-urban locations where the cost of providing broadband connections is too high to justify the investment on behalf of the telecom companies (Atkinson et al. 2008).

By contrast, if one looks at the most information networked economies, they tend to be associated with a high concentration of relatively dense cities and actively interventionist national and local governments that have grasped the importance of the information network infrastructure economy as essential to their global competitiveness. Nowhere is this more evident than in East Asia where broadband 'fibre to the home' (FTTH) accounts for 36 per cent of subscriptions in Japan and 31 per cent in South Korea as compared to an OECD average of 8 per cent in 2008.

Fibre or 'glass' as it is better known to industry insiders, is a much faster and more versatile material for broadband communications than the copper wire DSL connections that are common to most domestic and business telecommunication networks. In 2007, South Korea's embrace of advanced broadband technologies at the level of the general population was almost total – with a household broadband penetration of 93 per cent. Of the European countries, only Iceland came close, with 83 per cent of the population enjoying home broadband access. But given that half of Iceland's modest population of 300,000 inhabitants live in the capital city of Rekjavik, it is perhaps a less surprising discovery than the South Korean example (Atkinson et al. 2008: 8).

According to Sonn and Storper (2003), 'contemporary information and communication technologies have permitted a much more flexible and rapid interaction between product and process change in many industries, necessitating constant innovation and introducing new "islands of complexity" in the economic process which require dense, person-to-person communication among potential innovators'. As McLuhan and Fiore observe:

> Electric circuitry profoundly involves men with one another. Information pours upon us, instantaneously and continuously. . . : Our electrically-configured world has forced us to move from the habit of data classification to the mode of pattern recognition. We can no longer build serially, block-by-block, step-by-step, because instant communication insures that all factors of the environment and of experience co-exist in a state of active interplay.
>
> (McLuhan and Fiore 1967: 63).

This complex inter-layering of information and communication processing and distribution capacities in the service of what Thrift and French (2002) have described as 'knowing capitalism' requires the existence of 'creative clusters' (see Chapter 6) that bring together highly skilled and educated engineers, programmers, developers, writers, producers and editors along with the promoters and guardians of intellectual property (patent agents, lawyers, marketers) (Sassen 2001) all of whom need to be provisioned with a networked infrastructure that concentrates and intensifies differentiated access to physical space *and* the proliferating virtual space of privileged information flows (Graham and Marvin 2001).

Far from being a spontaneous crystal-like growth, 'the information city' can only be fully understood in terms of its relationship to the state at a variety of scales and according to the regulatory and governmental regimes that apply in diverse manifestations at the urban interstices of the globalising knowledge economy. The gathering and dissemination of 'media' is thus not simply an activity restricted to press, broadcasting and Internet-based information providers – whether they be private sector or citizen-centred organisations – it is intrinsic to how state and commercial actors code, organise and classify discrete population groups using an array of data gathering and analysis systems that date back to the earliest cities.

Biopolitics, code space and urban informatics

According to the second edition of the *Oxford English Dictionary*, the earliest known occurrence of the term 'statistics' dates from the publication of Helenus Politanus' satirical work *Microscopium Statisticum* in 1672, where its use pertained to 'statists or state craft'. Subsequent references to the term in the eighteenth century by German and French scholars continued to derive their sense of the term from the Latin word *status* or state. Hence the title of G. Achenwall's treatise of 1748, *Vorbereitung zur Staatswissenschaft*, which used the word *statistik* to describe how this new field could be translated as 'Towards a Political Science' (political science being the modern sense of the German term *Staatswissenschaft* or, literally, 'state-knowledge' – knowledge acquired by and for the purposes of the state).

Enumeration, classification and registration – indeed all forms of counting and recording the status of subjects within a given jurisdiction as we saw in Chapter 2 – are forms of governmentality associated with even the most ancient urban civilisations. The idea of 'population' and of measuring the size of discrete populations could not exist without the fact of the state, but this sense of the term 'statistic' has been lost through its routine application to that branch of mathematics concerned with numerical facts or data in general (Curtis 2001: 3). It was not until the nineteenth century that the exclusively quantitative aspects of 'the statistical sciences' as they began to be called were attributed to the study of numbers rather than the nature and process of government itself.

The collection of statistics has always been costly and cumbersome and is undertaken by government authorities essentially for two reasons: (1) to identify, locate and differentiate discrete populations, and (2) to levy taxes from which the expenditures of the state can be met. Before the nation-state was even a glimmer in the eye of Herder or Rousseau, city governors had established mechanisms through the adaption and extension of recording technologies to facilitate the first function, along with juridical and bureaucratic–coercive apparatuses to ensure the effective coordination of the second.

At the core of much socio-technological research is the rejection of the popular assumption that the relationship between human agency and technology is best viewed as a 'master–slave' dualism whereby information and communication technologies exist as tools for social interaction, consumption and production thus liberating the human subject from routine chores and investing him or her with ever greater degrees of choice and personal autonomy. In seeking to explain why this relationship is far more mutually contingent and conditioning than it outwardly appears, Graham and Murakami Wood draw on the work of actor-network theorists such as Bruno Latour in order to develop the notion of the 'hybrid collective' (Callon 1987; Law 1999; Bowker and Star 2000; Latour 2005).

This notion can be understood as a constantly interacting set of relationships between humans, non-humans (other living beings, natural processes, etc.) and 'in-humans' (human-made objects, materials, texts, etc.) where the non- and in-human elements 'are not merely passive in the sense of being imbued with value by society; rather they carry, change and produce power and value in a symmetrical relationship with individuals and groups of human beings'. Furthermore, these actor-networks or collectives are not amorphous or uniform but display distinctive properties including degrees of boundedness and territoriality. It is the way in which networks coordinate and align social and economic process and behaviours that determines the manner in which 'static categories' are created from 'a fluid reality'. In this way, 'hybrid collectives attempt to define and/or further expand their own boundaries through the hiding or erasure of alternate possibilities through categorical work, that is, the creation of both discursive and material boundaries' (Murakami Wood and Graham 2006: 180). It is this quality of hybrid collectives that we can observe most easily in the urban context through the deployment of code in the service of differentiated mobilities (Murakami Wood and Graham 2006: 179).

Surveillance, sorting and stratification in the networked society

The monitoring of individuals and groups (surveillance) is used for the purpose of categorising and differentiating populations (social sorting), and the stratification of populations that is thereby achieved relies on information and communication technologies and capabilities that evolve from and find their maximum expression in the urban sphere. As David Lyon writes, '[i]t is not merely that some kinds of surveillance seem invasive or intrusive, but rather that social relations and social power are organized in part through surveillance strategies' (Lyon 2007: 450). But, warns Kevin Haggerty, '[i]t has become profoundly difficult to say anything about surveillance that is generally true across all, or even most, instances' (Haggerty 2006: 39). This is not to refute Lyon's argument so much as to argue that the forms and purposes of surveillance are contingent on their time and place as well as the technologies and motives of control that lay behind their deployment.

As McLuhan makes clear, different technological media give rise to new forms of social ontology. Thus if the panopticon – whose command-and-control capacity relied on an indeterminable and ubiquitous gaze – was the dominant surveillant model in the nineteenth and much of the twentieth centuries, the 'new surveillance' that we encounter in the digitally controlled information and communication systems of cities in both the developed and the developing world has facilitated 'a step change in the power, intensity and scope of surveillance' (Graham and Wood 2007: 218).

The notion of 'urban splintering' in relation to infrastructural networks derives from Graham and Marvin's pioneering study (Graham and Marvin 2001), which has since been taken up and applied to a number of specific urban contexts (see also the following chapter of this volume). In the contemporary neo-liberal city Graham and Murakami Wood see a 'tendency towards technological lock-in, which threatens to divide contemporary societies more decisively into high-speed, high mobility and connected and low-speed, low-mobility and disconnected classes' (Murakami Wood and Graham 2006).

A similar if differently manifested digital divide exists in the Chinese city where, according to Qiu, '[t]he emergence of the information have-less [defined as those with limited incomes and limited influence in the policy process but who have begun to go online and use basic wireless phones] has become the definitive feature of informational stratification in Chinese cities'. The information have-less make up the vast majority of China's urban population. They comprise students and young people (15–24), migrant workers, those who have been made redundant, the retired and small-business people who earn the equivalent of $160–220 per month (Qiu 2009: 4–5). Largely due to the exponential growth in demand from this poor but communication-hungry urban population, China's mobile phone subscriptions reached 547.3 million by the end of 2007, while a further 84.5 million owned the Chinese 'poor man's mobile phone' known as 'Little Smart' (Qiu 2009: 53).

Qiu sees the Chinese *chengzhongcun* (urban village) and *danwei* (the residential work unit) as urban places that are not only essential to the urban have-less but, he argues, they 'would lose much of their distinctiveness without working-class ICTs [Information and communication technologies]' (Qiu 2009: 158). Because family and social ties are much more extensive and distanciated, as a consequence of China's rapid industrialisation and the influx of millions of rural *hukou* (officially registered) workers into its towns and cities, new migrant enclaves are appearing that rely on cheap and basic ICTs to share news of work opportunities, to maintain intimate contact with friends and family and even – where circumstances allow – to communicate and organise acts of resistance (Qiu 2009: 243).

Nowhere has the use of mobile communication devices and social networking software as instruments of urban-based social resistance been more apparent in recent times than in the case of Iran, where the disputed election victory of Mahmoud Ahmadinejad in June 2009 gave rise to protracted street protests in all of Iran's major cities. Supporters of Ahmadinejad's opponent, the former Prime Minister and reform-oriented Mir-Hossein Mousavi, took advantage of mobile phone-based micro-blogging and video utilities such as Twitter and YouTube to disseminate news of the protests and attacks on demonstrators by the police and government supporters.

So fundamental has been the use of social networking technology to the growth and identity of the protest movements in Iran that a number of commentators have referred to the phenomenon as 'Iran's Twitter Revolution'.[45] Indeed the media analyst Clay Shirky claimed that '[Iran's] is the first revolution that has been catapulted onto a global stage and transformed by social media' (Morozov 2009: 10).

Undoubtedly, the prevalence of affordable and widely available video and mobile phone equipment in urban population centres has made it much harder to suppress dissident opinions and to cut off news of repression from the outside world. But more sceptical voices point to the opportunities that social network surveillance provides to the authorities who are able to track posts and identify networks of friends and supporters for the purposes of repression. Because social media platforms such as Twitter are open to anyone to use, agents of the state can also masquerade as ordinary citizens in order to spread false rumours, denigrate leading activists and to accuse the movement of being controlled by foreign forces. Indeed the alacrity with which global media outlets such as CNN, the BBC and influential blog sites such as the Huffington Post reposted or broadcast the most disturbing images of the attacks on demonstrators, together with the increasingly US-dominated tweets on Iran's 'green revolution', was taken as evidence by the Ahmadinejad regime of a coordinated campaign by western powers to destabilise and ultimately bring down the Islamic Republic of Iran.

The use of show trials and the mass arrest and imprisonment of even very senior and moderate government opponents in Iran underlines the fact that where states continue to dispose of the monopoly of legitimate force, and where a sufficient consensus persists (as it appears to for Ahmadinejad in the more rural and smaller population centres in Iran), then talk of a Twitter revolution is certainly premature. Nevertheless, the Iranian protests serve to highlight the fact that the nature and form of urban-based political dissent and its dissemination through the use of interactive social media represents an important counter-tendency in the accretion of communication and informational power by the state.

Conclusion

As Dodge and Kitchin (2004: 209) have argued, 'Code is the lifeblood of the network society, just as steam was at the start of the industrial age. Code, like steam, has the power to shape the material world; it is able to produce space.' We have already noted how in the construction of geodemographic classifica-tions, a particular type of code is involved in the contemporary production of space. But who is involved in constructing the code? For whom is the code constructed? By whose agency? Why and how? And what features of the changing and dynamic coded space feed back into the ways in which that code is subsequently constructed and produced? These are all essentially political questions that seek to break the seamless, invisible circuits of power and knowledge that literally encode our urban world, but the answers to them are only likely to be found in the back offices of commercial geodemographic organisations (Burrows and Gane 2006), in the algorithms of CCTV facial recognition and profiling software (Bowyer 2004), in the biometric data warehouses and among the myriad 'data doubles' that permit an increasingly pernicious grading and sorting of 'credit risks' and 'security threats' (Lyon 2008)' (see Figure 14).

This chapter has been concerned with examining cities not only as basing and reception centres for a small number of global media behemoths, but increasingly as sites for the informatisation of place and the surveillance of 'spatially fixed' actors. Lefebvre draws the distinction between *spatial practice* (akin to the continuous or routinised use of space), *representations of space*, which 'are tied to the relations of production and the "order" which those relations impose, and hence to knowledge, to signs, to codes, and to "frontal" relations', and *representational space*, 'embodying complex symbolisms, sometimes coded and sometimes not, linked to the clandestine or underground side of social life' (Lefebvre 1991: 33).

Figure 14. British Transport Police and London Metropolitan Police surveillance promotional campaign.

Here Lefebvre's observations seem very close to those of McLuhan in the observation that representation creates 'order' just as new forms of communication technology create new forms of social ontology. It is only meaningful to think about the city as an assemblage of discursive formations if, along with Lefebvre, we acknowledge the material bases of all forms of urban representation – from the *polis* to the Prado. It is the 'oscillation' between real (i.e. material) differences and their sublimation, displacement and mystification that becomes manifest in both the built form and social organisation of the city that I hope to demonstrate and analyse in the following chapter.

Further reading

Dwayne Winseck's article on media ownership and media markets (Winseck 2008) provides a contemporary overview of the domination of global media ownership and media markets by a small number of transnational corporate behemoths, while Baker (Baker 2002) examines media concentration and monopolisation in US towns and cities. Scott McQuire's (McQuire 2008) book is an original and interesting take on the 'media city' and its symbiotic relationship with the built environment.

Burrows and Beer's 'Rethinking Space: Urban Informatics and the Sociological Imagination' in Prior (Prior 2010), provides a helpful synthesis of research on the transformation of urban space through the use of social informatics. Dodge and Kitchin's *Code/Space: Software and everyday life* (Dodge and Kitchin 2010) while not exclusively focused on the urban/regional dimension of software mediated spatialities is an essential reference for anyone interested in this emerging field of urban research. Ari Berman's blog for *The Nation* on 'Iran's Twitter Revolution' can be found online at www.thenation.com.

On urban surveillance, David Lyon's essay on 'Surveillance, Power and Everyday Life' (Lyon 2007) provides a good introduction to the intrusive powers of the electronic panopticon.

9 The landscapes of urban power

Either you bring the water to L.A. or you bring L.A. to the water.

Noah Cross, *Chinatown*

Introduction

When we think of the city it is almost always as a panorama or skyline – whether it be Manhattan's dramatic gothic skyscrapers, or the fog-shrouded dome of St Paul's cathedral and the clock tower of Big Ben which provide the symbolic link between the world of politics and finance along the waterfront of the River Thames in London, or the majestic Christmas-bauble domes of the Kremlin set in splendid isolation at the end of Moscow's Red Square.

Each city's urban landscape is unique from a strictly formal viewpoint, but as we discussed in Chapter 2 there are certain essential features of every city that can be attributed to the ways in which natural features, resources and conditions are harnessed and controlled. Every building wherever it is constructed relies on materials that have been either quarried or excavated or pumped out of the ground, which is to say that the transformation of nature and natural resources is intrinsic to the urban landscape. Once clay is transformed into bricks – or sand, water and chalk made into cement – the material essence of the city becomes an amalgamation of commodities, some of which are visible and some of which are not.

In Chapter 2 we discovered that even the earliest cities were associated with some aspect of intentional design and the reproduction of architectural features and decoration was commonly found in non-contiguous settlements that were peopled by the same clan or ethnos. Until very recently, the consensus among urban historians was that very few towns until the modern era were 'planned', and that most urban growth from the Middle Ages onwards was 'organic'. However, more recent scholarship has begun to dispute this view leading one author to comment that 'most if not all English towns and villages were planned . . . in the sense of being shaped by the conscious decisions of individuals or corporate bodies' (Palliser 2006: 9–10). The advent of modernity made this process of urban design at the same time both more rational and the response to it more unconsciously irrational as the experience of the industrial metropolis began to induce among an influential group of artists and intellectuals a romantic obsession with the idea of the city as an organic, corporeal and brooding presence – the Gotham City of twentieth-century adolescence twinned with the unfinished heresy of Gaudí's sacred towers and Alfred Hitchcock's *Vertigo*.

Power, ideology and urban form

The spatial organisation of settlements, whether planned or not, offers clues to the configuration of social and economic power within a given community. In trying to understand the different forces that gave rise to 'the production of space' the French urbanist and philosopher Henri Lefebvre wanted to draw attention to the ways in which the urban landscape has and continues to be the site of contestation across a range of fields from the symbolic to the physical enclosure and control of urban space by economically dominant interests and the legal–rational forms of control that accompany any given mode of production. Foucault sees a movement in architecture from the end of the eighteenth century towards a concern with

> problems of population, health and the urban question. Previously, the art of building corresponded to the need to make power, divinity and might manifest. The palace and the church were the great architectural forms, along with the stronghold. Architecture manifested might, the Sovereign, God. Its development was for long centred on these requirements. Then, late in the eighteenth century, new problems emerge: it becomes a question of using the disposition of space for economico-political ends.
>
> (Foucault 1980: 148)

What Foucault appears to be arguing here is that architecture constitutes a special kind of discourse, or rather special kinds of *discursive spaces*, which are implicated in strategies of power. There is a need, Foucault argues for a history

> of *spaces* which would at the same time be the history of *powers* (both these terms in the plural) – from the great strategies of geo-politics to the little tactics of the habitat, institutional architecture from the classroom to the design of hospitals, passing via economic and political installations.
>
> (Foucault 1980: 149)

Jeremy Bentham's celebration of his 'Panopticon' as moral architecture suited to both correction and the compulsory concentration of the poor in 1791 makes no bones about the multi-functional utility of surveillant space:

> Morals reformed – health preserved – industry invigorated – instruction diffused – public burdens lightened – economy seated, as it were, on a rock – the Gordian knot of poor laws not cut but untied, all by a simple idea in Architecture.
>
> (Bentham cited in Sicher 2003: 296)

What Angel Rama calls the 'lettered city' was also a key motif in the colonial subjugation of the Americas, which despite the existence of ancient settlements through the peninsular was seen by the Spanish colonists and missionaries as an urban 'tabula rasa'. Freed from the 'organic' disorder of the medieval Iberian cities the *conquistedores* 'adapted themselves to a frankly rationalizing vision of an urban future, one that ordained a planned and repetitive urban landscape and also required that its inhabitants be organized to meet increasingly stringent requirements of colonization, administration, commerce, defense, and religion' (Rama 1996: 1). Although the Church and especially the Jesuits held a strict monopoly in the souls of both settlers and *indios* alike in Ibero-America, the 'men of letters' who ordered the Latin American city were largely drawn from the urban bourgeoisie and as such there values were essentially Platonic.

The ideas of *The Republic*, revived by Renaissance humanism, arrived in America through the same Neoplatonist cultural channels that guided the advance of Iberian capitalism. And with Neoplatonic idealism came the influence of the quasi-mythical Hippodamus, Greek father of the ideal city – especially his 'confidence that the processes of reason could impose measure and order on every human activity' (Rama 1996: 2).

Rama sees this period in the sixteenth- and seventeenth-century European colonisation of South America, following Michel Foucault, as a moment when 'words began to separate from things, and people's understanding of epistemology changed from one of triadic conjuncture to the binary relationship expressed in the *Logique* of Port Royal, published in 1662, theorising the independence of the 'order of signs' (Rama 1996: 3).

However, it should not be assumed from this reading of Foucault that the 'order of signs' only assumed importance at the time of the seventeenth century. The Hippodamic city which the Greeks made the template of the Alexandrian urban system, was adopted in most of its aspects by the Romans, for whom the perpendicular layout of streets and thoroughfares was not only an efficient and effective method of planning new settlements, but conveyed in Bourdieu's terms a language of symbolic power. The fact that then as now the urban landscape remained largely invisible to those that experienced it on a day-to-day basis does not diminish the material impact of its symbolic ordering. Or, as Bourdieu would have it,

> we have to be able to discover [power] in places where it is least visible, where it is most completely misrecognized – and thus, in fact, recognized. For symbolic power is that invisible power which can be exercised only with the complicity of those who do not want to know that they are subject to it or even that they themselves exercise it.
>
> (Bourdieu and Thompson 1991:94)

In a similar vein, Henri Lefebvre writes:

> The members of archaic societies obey social norms without knowing it – that is to say, without recognising those norms as such. Rather they live them spatially: they are not ignorant of them, they do not misapprehend them, but they experience them immediately.
>
> (Lefebvre 1991: 230)

Finally here is a small passage from Walter Benjamin's short essay on the 'Mimetic Faculty':

> To read what was never written. Such reading is the most ancient: reading before all languages, from the entrails, the stars, or dance. Later the mediating link of a new kind of reading, of runes and hieroglyphs, came into use. It seems fair to suppose that these were the stages by which the mimetic gift, which was once the foundation of occult practices, gained admittance to writing and language.
>
> (Benjamin 1979: 162–3)

In other words, the production and interpretation of symbols and signs has always been a central feature of how society (and therefore the city) makes sense of itself and imbues itself with meaning and authority. Michel de Certeau reminds us that in contemporary Athens public transport vehicles are called *metaphorai* – one literally takes a metaphor to work (De Certeau 1984: 115). In just the same fashion, narrative serves this function of communicating the past structures of feeling that give rise to the urban complex *genealogically* to those that will inherit its future.

Planning and the imperial gaze

One of the most impressive of Imperial Rome's surviving architectural monuments is Trajan's Column. It was initially commissioned in 113 CE by the Senate to depict the army's successful conquest of Dacia (now mostly including the territories of present-day Romania and Moldova), which most sources agree was based on Trajan's own account of the campaign known as the *Dacica* – only a few words of which have survived. Unlike his predecessor Domitian, whose first campaign against the Dacians ended in disaster, and the second in what was considered to be an unfavourable truce in the eyes of most Romans, Trajan's two campaigns in 101–2 and 105–6 CE resulted in the conquest and then the complete subjugation of Dacia to the status of a client state.

The column is set upon a large plinth (which subsequently became Trajan's mausoleum after his death in 117 CE) and rises to a height of over 30 m. Its summit can be reached only by means of an unlit internal staircase. The outside of the column is decorated by a spiralling frieze depicting scenes from the Dacian wars in exacting detail (see Figure 15), which soon become a blur to the naked eye after the first few circumlocutions of the monument. A statue of Trajan himself was originally thought to have stood on a dome at the summit of the column, later to be substituted by the figure of St Peter – a practice often adopted by the Church in order to provide high-profile settings for the Holy Family and the saints.

The exact purpose of Trajan's Column has divided historians for many years, since it seems perverse to have devoted such craft, skill and labour to a monumental text that is all but 'unreadable' to the average passer-by, and upon whose precise meaning even specialists on the period can only speculate. According to Penelope Davies, however, the key to understanding the monument is the circumambulation that – wittingly or unwittingly – visitors would have performed as 'a perpetual reenactment of the funerary ritual' (Davies 1997: 56).

Another function of the monument, the ground for which had to be excavated from the mount that now forms the Quirinale, was to provide a viewing position or *belvedere* from which to admire the grandeur of the imperial city, and most importantly Trajan's own magnificent forum within which the column was set with its market, law courts and library. The inscription on the capital of the column not only lauds Trajan's divine ancestry and six appointments as emperor and consul, but also the achievement of the Roman architects and engineers who succeeded in removing an entire hillside in order to create the *campo* on which Trajan's forum now stands.

Apollodorus' layout of the forum (see Figure 16), which would have been immediately obvious to a military officer, was based – according to Rodenwaldt – on the military camp, possibly that of the Praetorium at Vetera, with the column taking the place of the army's standards (Davies 1997: 61).

The timing of the construction of the column at a period when Trajan's armies were facing military defeat in north Africa could not have been accidental given the external and internal threats that the imperial city faced at the time. The

Figure 15. Trajan's Column showing part of the spiralling frieze depicting the Dacian wars. © Ariy Zimin/Dreamstime.com

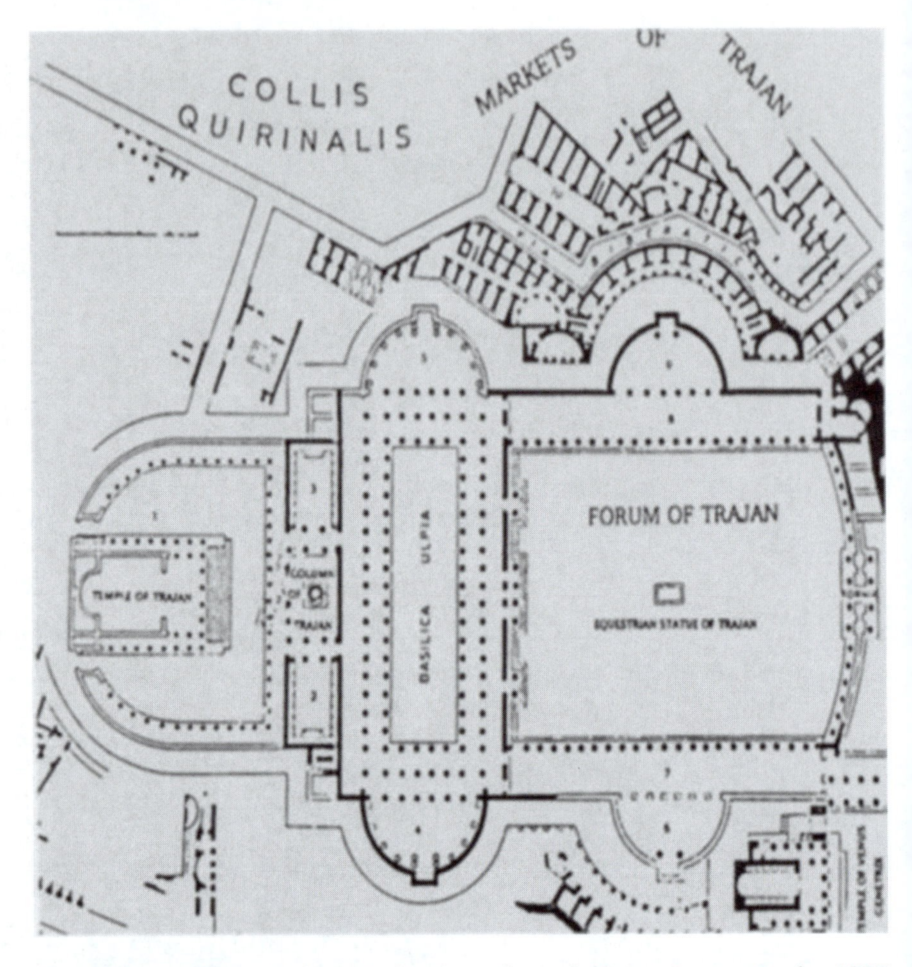

Figure 16. Apollodorus's layout of Trajan's forum in Rome. Source: Davies 1997 © Archaeological Institute of America

column served thus not only to promote the prospects for future wars but also to restore the flagging morale of an unsettled populace by symbolically evoking past triumphs through the invocation of what Lefebvre describes as 'the lyrical space of legend and myth, of forests, lakes and oceans', which, '[vie] with the bureaucratic and political space to which the nation states have been giving form since the seventeenth century' (Lefebvre 1991: 231).

This use of the monumental to evoke the eternal recurrence of the aggrandising state is an abiding theme in the patriotic architecture of the metropolis, but the structuring of the urban landscape around the imperial gaze is not limited to the formal splendour of Berlin's Brandenburg Gate or Napoleon's Arc de Triomphe in Paris, or the Washington Memorial, but rather each of these sites exist as axes of symbolic power around which are topographically ordered not only the capital city but the entire geography of its national and colonial hinterlands.

State power and defensive urbanism

The point at which city authorities became less concerned with foreign enemies and more attentive to potential enemies within is hard to define precisely, but it certainly coincided with the period that saw the removal of city walls and the emergence of a permanent wage-dependent workforce in the towns and cities that were the first to industrialise in the late eighteenth and early nineteenth centuries. As Lewis Mumford puts it in characteristically bold terms:

> Protection gave way to ruthless exploitation: instead of security, men sought adventurous expansion and conquest. And the proletariat at home was subject to a form of government no less ruthless and autocratic than that which ground the barbaric civilizations of North and South America into pulp.
>
> (Mumford 1989: 361)

The industrial revolution required an unprecedented concentration of wage workers in the towns and cities not only in the European and North American metropoles, but in the burgeoning cities of empire – from Calcutta to Shanghai and from Cape Town to Kingston. The 1789 revolution in France and its overseas colonies had encouraged urban-based revolutionary nationalist movements in other parts of Europe in the early nineteenth century and in Latin America where as Chapter 4 noted the emergence of an indigenous middle class was beginning to use its identity as an educated, urban-based society to challenge the sovereignty claims of its colonial masters. Urban-based alliances between proletarians and bourgeois and settlers and natives began to shake the authority and confidence of domestic and colonial sovereigns who came to realise that the city – as well as being the source of wealth and prestige – was at the same time a harbinger of revolt and new political imaginaries.

Craig Calhoun argues that, following the first wave of revolutions in the nineteenth century, state elites realised that containing all their institutional power in the national capital made them vulnerable to being overthrown by a local mass uprising. However, by extending the state administration (in particular the repressive apparatus of the state) throughout the country 'the risk of an urban insurrection becoming a true revolution was sharply reduced'. Indeed the Paris commune of 1871 demonstrated how the state could continue to survive and triumph despite losing control of the national capital. As Calhoun observes, after 1848 with the advent of modern communications technologies (railroads, telegraphs, improved forms of administration) in the economically advanced countries, 'no modern European (or, more broadly, "rich country") government could be toppled simply by riots in a capital city' (Calhoun 1998: 376).

The displacement and territorial distanciation of the state apparatus did not, however, imply an abandonment of strategic urban space to 'the mob'. Rather, the systematic planning (or better, 'creative destruction') of the city assumed an even greater importance as the superintendents of *ancien régime* hegemony deployed the recently developed science of town planning (what the situationist Guy Debord described as 'the science of the state')[46] in the service of a newly assertive and moneyed bourgeoisie through a new governmentality of space. As David Harvey explains, this practice was particularly advanced in Second Empire Paris where, '[t]he building of new boulevards . . . was considered strategic, designed to permit free lines of fire and to bypass the hard to assail barricades erected in narrow, tortuous streets that had made the military suppression of 1848 so difficult' (Harvey 2006: 20).

However, military security was not the only purpose of the demolition and boulevardisation of large swathes of central Paris. These public investments in a new urban form, intended to facilitate the movement and leisure-based consumption of the post-republican Parisian bourgeoisie, were an early example of deficit-financed 'regeneration', which Harvey describes as 'some mix of civilian and military Keynesianism' (ibid. 21). In order to achieve the rapid appreciation of property values that Haussmann's reconstruction of the capital was intended to achieve, not only did the physical and built environment of the central city need to succumb to *embourgeoisement*, but the social and economic habitus that had given the central *arrondissements* such a fearful reputation for insurrection had to be reconfigured through the suppression of local industries deemed offensive to middle-class eyes, ears and noses and by the physical expulsion of its workers to a safe distance from the heart of government and commerce.

This fear of the crowd was largely expressed by governments and city administrators in the nineteenth century, but by the second half of the twentieth century, as Nan Ellin observes, 'defensive urbanism' had become incorporated into urban design practices at a much more molecular level, reflecting 'an attempt to satisfy longings for community and security' (Ellin 1999: 1). As the preoccupation of wealthy urbanites shifted from the enemy at the gates to 'the enemy within' proposed solutions to the problem of crime and disorder turned to technologies of what might be called hard and soft forms of social sorting.

Hard social sorting involved the further development of Haussmann-style physical destruction and reconstruction of neighbourhoods with the intention of displacing, isolating and containing the city's poor populations. But this was not only achieved by the use of residential relocation through slum clearance programmes, but just as importantly through the expansion of commercial and business zones, high-speed transport networks, health and education provision, and the placement and concentration of law enforcement personnel and facilities (Berman 1982; Harvey 1989b, 1989c; Davis 1992; Graham and Marvin 2001; Flusty 2005).

As the city changed from a space of socially and economically diverse but adjacent populations who were often to be found living in the same street and at times even in the same building to a more functionally and spatially segregated city that both represented and exaggerated the socio-spatial divisions present within society as a whole (Parker 2000), there was a concomitant increase in the development and deployment of planning instruments as a key tool for the categorisation and segregation of both urban functions and urban populations. Matthew Gandy refers to this phase in the modernisation of Paris in the 1850s and 1860s as being 'predicated on a holistic conception of the relationship between the body and the city, which drew on a series of organic analogies to compare the new city with a healthy human body' (Gandy 1999: 24).

The underground galleries that carried away human effluent and replenished the city with fresh water were meant to function as the hidden organs of the metropolitan organism. The construction of the new Parisian sewage system by Haussmann's chief engineer Eugène Belgrand was partly a response to the cholera epidemics of the 1830s and 1840s and the economic and social crisis that a city of over a million inhabitants faced when reliant on medieval methods for obtaining drinking water and the removal of human waste. But at the same time Haussmann's public works scheme allowed the 'underused capital and labour behind the economic depression and political violence of 1848 . . . to be channelled into the reconstruction of the built environment through a deficit-financed economic strategy, rooted in Saint-Simonian ideas' (the proto-Keynesianism identified by Harvey) (Gandy 1999: 28). Through the propagandistic employment of photographic artists such as Félix Nadar, the representation of subterranean architecture's uncanny aura (and its inevitable association with the regions of the spirit and the unconscious) sought to reinforce 'the ambiguous role of modern technologies in providing an illusion of complete control and comprehension of complex urban societies' (Gandy, 1999: 26).

Networked infrastructures and the resources of urban power

The achievement of the Irish-born emigrant and self-taught water engineer William Mulholland was no less significant than that of Belgrand, but his audacious plan for the construction of a Los Angeles aqueduct was not simply

Figure 17. An aqueduct in outer Los Angeles supplies precious water to the ever-thirsty city. © Ron Chapple Studios/Dreamstime.com

an engineering one. The diversion of water (and hence the opportunity to attract population) from the settlements along the watershed of Owens Lake successfully escaped the protests and dynamitings of the angry upstream residents and the obstacles of state and federal horse-trading, log-rolling and corruption made it possible for a fringe, coastal settlement on the edge of the Pacific Ocean which was home to some 44 settlers in 1781 to become one of the world's richest world cities with a population in the greater metropolitan region approaching 18 million people by the twenty-first century (see Figure 17).

The story of Mulholland's progress from a penniless ditch digger to founding father of modern Los Angeles is a classic American story of personal ambition, guile, determination and the vagaries of fortune. The collapse of the St Francis dam in 1928 which killed an estimated 450 people ruined Mulholland's reputation as a civic-minded genius and inspired a number of literary and cinematic accounts of Californian corruption – among them Roman Polanski's *Chinatown*[47] – in which the battle for control over water resources overshadowed even the bitter conflict between labour unions and employers in the region's booming manufacturing, construction and agricultural industries in the inter-war years (Davis 1992).

As Kaika and Swyngedouw write:

> When the urban became constructed as agglomerated use values that turned the city into a theatre of accumulation and economic growth, urban networks became the iconic embodiments of and shrines to a technologically scripted image and practice of progress. Once completed, the networks became buried underground, invisible, rendered banal and relegated to an apparently marginal, subterranean urban underworld.
>
> (Kaika and Swyngedouw 2000: 121)

There was a *braggadocio* surrounding the great infrastructural projects of nineteenth-century Europe and their colonial empires that carried through to the building of 'cities in the wilderness' such as Chicago and Los Angeles in the 'New World'. Kaika and Swyngedouw identify the ways in which infrastructure networks became 'urban fetishes' in the earliest phase of modernity, subsequently developing into a material and cultural rallying point for the 'ideology of progress'. Only as ideologies of progress begin to fail and when nature reasserts its domain over the 'new Babylons' of Chicago in 1871, San Francisco in 1909 and Los Angeles in 1928 do we begin to witness the banalisation and backgrounding of networked infrastructure, as modernity – along with its accompanying political discourse – becomes recast in new ways.

The routinisation of the networks of urban power is part of the disenchantment of modernity that Max Weber also associated with the increasingly corporate and bureaucratic nature of organised politics. Planning, the securing of energy and water resources, the construction of schools, hospitals, offices, factories and houses does not diminish in importance in late modernity, but rather these 'hard wear' functions of the city become absorbed within the autopoeitic system of the territorial state (Jessop 2001), only to be exposed in their full ideological antinomies at moments of spectacular crisis – such as 'natural disasters'.

Othering the storm: the urban political ecology of Hurricane Katrina

According to the United States Congressional Research Service, the hurricane that made landfall on the Gulf Coast states of Louisiana and Alabama on 29 August 2005 affected 5.8 million people within the storm area and contributed to the deaths of some 1,000 people within the three states affected. Despite being widely reported as a 'natural disaster' and an extreme weather event whose consequences could not have been foreseen or planned for, Ivor van Heerden, a hurricane expert at Louisiana State University in Baton Rouge described Hurricane Katrina as 'a preventable disaster'.[48] Neither was the levee system 'overwhelmed' by floodwater – the 17th Street levee collapsed while the water level was half a metre below the top of the levee wall and its failure was almost certainly due to inadequate reinforcement of the foundations given New Orleans' notoriously swampy soil conditions.[49]

Few commentators accepted that race and space were relevant factors in the disaster, reinforcing a widely articulated view by politicians and the mass media that advance warnings were given and that those who were left behind to face the consequences of the storm had chosen to stay against official advice – they therefore had no one to blame but themselves (Powell et al. 2006: 64). A closer analysis of the distribution of New Orleans low-income residents reveals a very different picture, however. The New Orleans poverty rate according to the 2000 census was 28 per cent (more than double the national average) and the city became progressively poorer during the 1980s and 1990s with an increase in the number of census tracts in which at least 40 per cent of the population are designated as poor increasing from 30 in 1980 to 49 in 2000. Overwhelmingly, the poor of New Orleans are African American and Latino, with the African American poverty rate at 35 per cent, amounting to three times the rate for the city's white population. According to the Index of Dissimilarity and Isolation (which measures the degree of ethnic segregation or integration) New Orleans is one of the most racially segregated of America's 50 largest metropolitan areas: 43 per cent of poor African Americans live in the poorest wards, and these wards are nearly all in the part of the city that lies below sea level. The Lower Ninth Ward which experienced the worst of the flooding had a 98 per cent African American population (Hartman and Squires 2006: 3 and 64).

As Powell et al. argue, 'disasters[,] whether "natural" or man-made, disproportionately affect the most marginalized members of society' (Powell et al. 2006), but the disproportionately of risk is not only a function of socio-economic and ethnic status – it is also a function of the way in which exposure to risk is mitigated or aggravated in discrete urban locales as a consequence of the human-made environment. When the Mississippi River burst its banks in 1927 threatening to overwhelm the much less extensive flood defences of New Orleans at the time, the city authorities decided to dynamite a breach in the levee, thus inundating the unfortunate inhabitants of down-river St Bernard Parish (Barry 1997) which was to suffer the same fate nearly 80 years on. While in the 1960s, the US Army Corps of Engineers (USACE) – which along with the Federal Emergency Management Agency (FEMA) was roundly criticised for the Katrina disaster – proposed building a flood wall that would have protected both the downtown and the city's poorer neighbourhoods. However, the USACE plan was shelved because 'residents' (code for the city's more affluent tax-paying voters) and officials objected on cost and 'aesthetic' grounds.

When Hurricane Hannibal battered the city in 1993, the floodwall had been diverted at great expense to protect Mark Twain's birthplace, while the rest of the South Side's poor community found their homes submerged under 2.5 m of water (Steinberg 2006: 19). By the time Hurricane Katrina struck 12 years later, nothing had been done to reduce the physical and socio-economic vulnerability of the New Orleans poor who remained crowded into the below-sea-level neighbourhoods on either side of the Y known as 'the funnel' that borders Lake Borgne, the Industrial Canal and the Mississippi River. The populations in these most flood prone neighbourhoods, many of whom were more recently arrived Latino and Hispanic residents rather than the longer-established African American community,[50] were the least likely to be able to evacuate voluntarily due to their lack of mobility (51,000 New Orleans residents – 27 per cent of the adult population – did not own a car), lack of insurance (few home owners in the most flood prone wards could obtain or pay for home and contents insurance), and the dependence on welfare payments or salary cheques that have to be physically collected. Combined with a general scepticism towards the pronouncements of city, state and federal officials this induced what for some was a fatal symbiosis between the storm-threatened streets of New Orleans' most deprived neighbourhoods and its least affluent residents (Cutter and Emrich 2006: 105).

The idea that disasters such as Hurricane Katrina could have been better handled if New Orleans had the benefit of an urban regime involving 'a stable and long-lasting partnership of public and private resource providers' underscores the prevailing consensus in mainstream urban political studies that structural and environmental factors are essentially backdrops to the public policy 'problem solving' activities of urban stakeholders and government agencies. The differential impact of urban trauma is thereby reduced to a problem of governance, the solution to which is the development of a regime-type 'shared agenda' or better inter-agency coordination and 'mission focus' (Schneider 2005; Burns and Thomas 2006).

In this literature there is far less emphasis on the repeated identification of New Orleans' ethnic minority poor by political and media elites as *the* problem, for which Hurricane Katrina was seen as either divine retribution or a heaven sent solution.[51] Housing and Urban Development (HUD) Secretary Alphonso Jackson was quoted as saying New Orleans 'is not going to be as black as it was for a long time, if ever again', while New Orleans City Council President Oliver Thomas insisted that, '[w]e don't need soap opera watchers all day', insisting that if displaced residents wished to return to public housing, 'they better want to work'.[52]

Figure 18. An abandoned boarded-up building in downtown New Orleans after Hurricane Katrina. © Pixelthat/Dreamstime.com

Suddenly, a 'natural disaster' had become a game of 'musical homes' where the number of affordable dwelling units had been reduced by act of God with enough supply for only the most worthy families to return (see Figure 18). The great storm, whose extreme effects had been intensified by the desolation of the coastal wetlands through petroleum extraction and the cutting of canals to allow greater amounts of ship traffic over several decades, literally uncovered a political ecology of vulnerability and risk that mainstream American society had refused to confront and still continued to deny even as the Superdome filled with almost exclusively black and Latino families (Dyson 2006).

Conclusion

The earliest cities were principally permanent shelters designed to protect their inhabitants from the adversities of the elements and the risks of the natural world. But, as cities became more complex and internally differentiated, their physical and organisational design reflected the structurations of power – political, economic, social and cultural – that inscribed the urban landscape from the landless labourer's cottage to Caesar's palace. Traditionally political scientists and political sociologists have paid little attention to the physical environment of the city, the design and economics of its infrastructure and the encoding of material discourses in the layout of streets, the deployment of public art and monuments and the barriers and filters that are aimed at the control of populations.

The explanation for this may lie in the strong bias within political and administrative science towards tangible data sources such as votes, party membership figures, budget allocations, policy outputs and so forth, rather than an engagement with the less tractable and historically sedimented *longue durée* of social differentiation that corresponds to the distinct but connected worlds of human, economic, social and symbolic capital (Bourdieu 1986).

The increasing disconnection between the public realm and the concept of the political is comparatively recent in origin, and as we observed in relation to the 'disappearance' of networked infrastructures, is closely tied to the emergence of the techno-bureaucratic stage of modernity in the urban sphere. The ancient Athenians had the advantage of living in a city that was also a nation in its own right, and therefore the correspondence between the physical space of the *agora* and the democratic community of the polity was well understood as a necessary and indeed vital one. In the contemporary city this dependency has not been entirely lost in the view of Zygmunt Bauman, who (drawing on the work of Cornelius Castoriadis) sees the *agora* as a politically articulated public space, which is nevertheless distinct from the state (the *ecclesia*) and from the private sphere of the household (the *oikos*). Nevertheless Bauman identifies late capitalism with the relentless subordination of the *agora* to the private regarding sphere of the *oikos* and thereby to the systematic depoliticisation of the profoundly urban qualities of civil society (Bauman 1999).

Certainly the shrinkage of the civil public sphere in cities around the world and the increasingly splintered urban habitat that hundreds of millions of urbanites are forced to live in is a very different idea of community to that envisaged by Plato, Aquinas and More (Pinder 2005). The production of space is the result of both the spatial production of commodities and the commodification of the urban landscape. It gives rise to a form of politics – as we saw in the tragically inadequate response of all levels of government in the wake of the Hurricane Katrina disaster – which is increasingly incapable of mitigating the dual threats of 'natural risk' and the global economic storm by resort to nature-defying technesis and an over-leveraged and hollowed-out post neo-liberal state.

Further reading

David Pinder's insightful study, *Visions of the City: Utopianism, power and politics in the twentieth century* (Pinder 2005) provides a wealth of detail and analysis on the centrality of utopianism to modernist urban visions and the ways in which the urban landscape both reflected and made possible a certain configuration of power in the cities of the twentieth century.

Maria Kaika's *City of Flows* (Kaika 2005) presents a fascinating analysis of how the making of modern Athens relied on the diversion and control of precious water supplies, and uses this case study to explain why a critical understanding of the control and use of natural resources in cities is essential to understanding the nature of the urban landscape and the distribution of vital natural resources.

Bullard and Wright's *Race, Place, and Environmental Justice after Hurricane Katrina* (Bullard and Wright 2009) presents a compelling account of how institutional flaws, poor planning and false assumptions about the lives and resources of New Orleans' poorest inhabitants all contributed to one of the worst urban disasters in contemporary American history.

Part V
Conclusion

10 Power and politics in the city

This book's main ambition was to broaden the discussion of urban power from its traditional focus on urban government and urban politics – while not neglecting the centrality of both categories to urban political life and urban studies in general – to a more diverse set of fields in which cities, regions and states operate as both the site and agent of a complex ensemble of governmentalities.

In order to unpick this often-tangled skein of power relations I have found it necessary to compartmentalise aspects of the urban experience that in reality are co-present. This has advantages in terms of focusing on a particular subject in a sufficient degree of detail, one hopes, to shed some light on the phenomenon, but one also risks losing sight of the bigger picture in which power is transacted in a million different ways in a million different spaces. A similar observation can be made about the category of social analysis represented by the term 'the urban' – a phenomenon whose ubiquitous development and increasing embrace of the population of the globe problematises a parsimonious distinction between city and country that classical urban theorists such as Weber and Simmel took for granted. At the same time 'cities' are morphologising into vast, sprawling slums with (to the outsider) no obvious trappings of the conventional city, while in the post-industrial West suburbanisation has given rise to ex-urbanisation, the growth in separate, secluded and increasingly gated communities and a concentration of poverty and deprivation in the old metropolitan cores.

In a recent essay, Robert Beauregard helpfully distinguishes between three modes of urban theorising: these he terms critical theory, heuristic theory and inter-textual theory. Critical theory (in its lower case sense) is associated with critical perspectives that are broadly affiliated to Marxist–structuralist understandings of historical and social development (also referred to as 'grand narrative' or 'meta theory'); heuristic theory involves empirically based 'bottom-up model building' that achieves (or fails to achieve) theoretical valency through imitation (or refutation) by other investigators; and inter-textual theory involves 'writing about writing' and is associated with the linguistic or 'cultural turn' in the social sciences that originated in literary and philosophical studies and which (in its post-structuralist and post-modern guise) is dismissive of positivist and 'scientific' accounts of human knowledge.

This volume has tried to represent all three of these traditions in urban studies research, but I think it would be fair to say that the study of urban politics in general (with the probable exception of the urban political economy community) tends to operate within the terms of reference of the heuristic model. This is an effective strategy for building 'mid-range' theory that offers the possibility of some comparative generalisability and – when it is done well – a robust empirical underpinning that can be subject to qualification and contradiction. However, what heuristic urban studies gains by a close-up knowledge of particular urban polities it loses in terms of a synoptic view of how urban political cultures evolve in space and time.

This is why so much of this volume has drawn on the work of historians, anthropologists and archaeologists who are capable of providing insights into our social world because they reveal in the otherness of cultures that are remote in time or 'development' that which is essential to our own. In making the extrinsic meaningful we have a fuller sense of what is common (intrinsic) to all human cultures, including those based within urban settlements. At the same time temporal–spatial comparisons allow us to isolate and assess the extent to which social, cultural, economic and political traits are dependently or independently variable according to context and conjuncture.

The trait of 'democracy', for example, reveals itself to be the result of multiple determinations – the least significant of which is its political character (if by political we mean its formal constitution as a mode of government). Democracy is not a static property of plural/pluralist societies but a dissemination of the nomadic idea of 'representation' in which the part stands in for the whole (synechdoche). For example, the material discourse of the republic (see Chapter 2) appears at different points and in different contexts in human history – but it can only function as a system of government where there is the compossibility of unity in difference, whereas autarchy and tyranny are consequences of the incompossibility of difference (Badiou 1999).[53] Difference can only be made compossible through movement. Typically movement – or nomadism (a term proposed by Deleuze and Guattari, 1983a) – occurs during periods of rapid economic and cultural change, for example with the consolidation of Athens as a major urban centre for agrarian surplus and trade during the Archaic period (750–480 BCE), in thirteenth- and fourteenth-century Italy when an emergent merchant class 'rediscovered' classical philosophy and reason and recruited it for the service of seigneurial republicanism, and at the time of the European enlightenment in the late eighteenth century which was accompanied by the 'proto-industrial' phase of capitalism. In this reading, democracy is less a feature of continuity in urban societies so much as a mode of governmentality associated with rupture and change (Foucault et al. 1991).

Another aim of this volume was not to lose sight of the fact that many urban populations still experience power as domination – or as violent 'kinetic' power as distinct from the 'molecular' power we increasingly associate with the seamless, invisible regulation and arrangement of populations in the cities of the Global North. Trying to understand why particular urban locations and certain urban populations experience power as *dominio* and others as *direzione* requires us to investigate, as Gramsci demonstrated, the historical conditions under which a definite type of politics is possible (Gramsci 1971). Thus, as Robert Putnam (1993) has shown, it mattered enormously that many of the cities of northern Italy, unlike their southern counterparts, could lay claim to hundreds of years of autonomous republican government when we attempt to explain the differential attitudes towards the *Rechtsstaat* and enthusiasm for the modern *agora* of civic associations and social networks in Emilia-Romagna as compared to Campania.

At the same time we should be wary of ascribing too much path dependency to the pre-history of contemporary urban polities, not least because the scales, interdependencies and modes of governance have changed so dramatically in recent decades. Some of this change is quite subtle and difficult to perceive if one focuses on cities with a strong democratic and 'statist' tradition – as in Nordic Europe – but even in countries such as Denmark the increasing presence of extra-European minorities and migrants has changed the terms of political discourse in a more conservative direction and in a way that has sought to distinguish 'tolerance' from a multiculturalism that an increasing number of Danish voters refuse to accept (Skidmore-Hess 2003).

In other cities such as those of the former Soviet bloc countries, the cities of the highly indebted and 'structurally adjusted' Global South and the 'maximum cities' that have experienced runaway growth and equally rapid urbanisation in the so-called BRICs (Brazil, Russia, India and China), the transformations have been so dramatic that history would appear to have nothing to teach us about how the rule-less politics of such urban spaces with their drug-mafias, epidemic corruption and booty capitalism came to exist.

The government of cities, in other words, increasingly depends on the management of social difference, but with the general decline of Keynesian welfare capitalism as a strong material discourse for urban redistribution and social mitigation, and the accompanying rise of neo-liberalism as a default strategy of urban governance, the imagineering of cities has accelerated the marginalisation and erasure of 'unmarketable publics' – from the homeless, to migrants, and religious and political extremists. In deference to urban taxpayers and consumers, civic public space has been made 'spiky' for these unwelcome urban denizens through attention-grabbing revanchist 'street clearances' and the increasing privatisation or de-publicanisation of the urban commons, as a number of critics have shown (Flusty 2004; Low and Smith 2006).

Because the city's powerless and marginal populations have to go somewhere, urban public authorities have implemented containment operations – these are places where new hetero(dis)topias, civic norms and laws are abated in the creation of 'zones of tolerance'. A contemporary version of *Gin Lane*, which has its fictional parallel in the so-called 'Hamsterdam' drug-containment initiative in the HBO crime series *The Wire* and a real-life equivalent in Vancouver's Downtown East Side, where the only legal safe injection site in North America continues to be strongly contested (Kerr et al. 2003).

Re-casting politics/re-thinking power

Urban theory has long wrestled with the highly complex relationship between power and the city – and workable definitions have often foundered on the ineluctable nature of 'power' itself. Existing definitions of urban power and indeed the vast majority of urban research tend to focus on processes of power or power relations *within* cities, but what is generally lacking, even in critical urban theory, is a fuller consideration of the systemic and *structurated* means by which, to paraphrase Simmel, the possibility of urban society is maintained (Simmel 1910); in other words, how the power of cities and of urban spaces in general is established, maintained, concentrated or dispersed.

In contrast, Peter Saunders seeks to move away from the idea of the city as having intrinsic properties by arguing that 'urban social theory cannot be constituted around the object of the city or the problem of space' (Saunders 1986: 289). In his view, cities are spaces where consumption happens but it is not the space that is important or determining, but the nature of the consumption and the social relations that it entails. The problem with this approach is that the same lack of parsimony that Saunders accuses social theorists such as Giddens and Harvey of (i.e. the failure to demonstrate the centrality of space to the process of social reproduction, or to relate the theory of capital accumulation and crisis to the question of spatial form) applies equally to the relationship between individual and collective forms of consumption and 'the urban' as an empirically and theoretically distinct sphere (Saunders 1986: 287 and 351). Ironically, David

Harvey would seem to concur with Saunders in denying that the city is anything other than a place where capitalism organises and reproduces itself. Harvey believes that

> the reification of cities when combined with a language that sees the urban process as an active rather than passive aspect of political-economic development poses acute dangers. It makes it seem as if 'cities' can be active agents when they are mere things.
>
> (Harvey 1989a: 5)

The fact that Saunders and Harvey both seek to challenge and refute the possibility of cities as collective agents should not surprise us, since Weber and Marx – to whom both authors owe their respective intellectual allegiance – despite having interesting things to say about cities and urbanisation, did not make a distinction between urban society and industrial society in general. By contrast, Le Galès and Bagnasco, in seeking to emphasise the continuing relevance of Weber to urban political sociology, argue that the diminishing importance of nation-states has created a 'power vacuum' that has provided new potential for local and regional action. Cities have been quick to seize the opportunity offered by the new state spaces created by global/regional rescaling and 'have become political and economic actors and increasingly created identities of their own' (Bagnasco and Le Galès 2000: 6 in Häussermann and Haila 2004: 52).

At the same time we need to be alive to the less conventional and orthodox manifestations of urban power – the material discourses of the urban complex that assume myriad forms: from the biometric code that may be used to identify and track the 'terrorist gait' (Goffredo et al. 2008), to the postcode-based algorithms that are used to divert urban police resources to crime-fear 'hot spots' (Graham 2010), to the public order offences that are used to criminalise female sex workers (but not their male clients) (McDowell 2009), to the increasingly globalised pattern of homophilic geodemographical sorting made possible by the bespoke data warehouses of location marketers (Savage et al. 2004, Burrows and Gane 2006). Far from space and power being distinct and non-dependent manifestations of urbanism – in the highly informatised world of global knowing capitalism (Thrift and French 2002) – the two have never been more cynically and insistently combined.

In arguing for a critically informed and searching hermeneutics of urban power it is worth revisiting the questions asked by Fredric Jameson in the context of the urban experience

> How do we pass . . . from one level of social life to another, from the psychological to the social, indeed from the social to the economic? What is the relationship of ideology, not to mention the work of art itself, to the more

> fundamental social and historical reality of groups in conflict, and how must the latter be understood if we are to be able to see cultural objects as social acts, at once disguised and transparent?
>
> (Jameson 1971: xiv)

This wonderfully resonant image of the city as a 'cultural object . . . at once disguised and transparent' inspires us to make the study of urban power a starting point rather than an end point of critical urban studies. It also encourages us to think through and across the processes of change and exchange that have transformed the nature of the urban experience in a constant narrative of flows, interruptions, fixings, ruptures and combinations.

The urban theorist must not only aspire to be a philosopher of the city but also to be a critic of urbanism – seeing the city as a work of many authors in which the socio-economic function, while fundamental, is not exclusively held to determine its polymorphous form (Parker 2000). In privileging the capitalist mode of production as a determining, permanent and animating force in the production and reproduction of contemporary urban society, urban theory risks obscuring the antinomies, contingencies and time–space specificity of the capitalist city, rather than seeing the late industrial metropolis and its increasingly disaggregated variants as but a further iteration in the *longue durée* of possible urban worlds.

As I hope to have shown, the multivalent character of 'the political' and the rhizomic nature of power in the context of advanced, densely populated human settlements requires us to go beyond a single determining narrative in search of the pattern language (Alexander et al. 1977) which binds the city's various material discourses together and that animates the urban with its enduring attraction for more than half the world's population.

In arguing that cities will continue to be a key subject of enquiry and analysis for students of political power, I hope to have shown that the focus of future research needs to shift away from a view of the city as essentially a 'power container' and more towards seeing the urban complex as a multi-dimensional, dynamic and self-reproducing socio-economic system that is structurally shaped by the development of local, regional and national modes of governance while remaining vulnerable to the destabilising impact of environmental threats, global flows of labour and capital, military and paramilitary conflict and organised crime and corruption. In other words, cities are not just the physical and territorial embodiment of trans-historical social, economic and cultural forces; they are the principal agents of transformation and human development in the modern world.

Notes

1 Following Talcott Parsons, I prefer to translate *Herrschaft* as 'authority' rather than 'domination' because the latter implies that power is essentially a zero sum function whereas Weber also intended *Herrschaft* to mean the power to organise and administer – or 'to make things happen' and this need not involve the subordination of an unwilling subject (Whimster 2007: 233).

2 Andrew Sherratt, 'Environmental Change: The evolution of Mesopotamia', www.archatlas.org/EnvironmentalChange/EnvironmentalChange.php.

3 *Dispositif* in French literally means 'device' but in Foucault's terminology *dispositif* is closer to an apparatus or a 'heterogeneous ensemble consisting of discourses, institutions, architectural forms, regulatory decisions, laws, administrative measures, scientific statements, philosophical, moral and philanthropic propositions' (Foucault 1980, 198–228).

4 N. Balestrini (2004). *Sandokan: storia di camorra*. Torino, Einaudi. Author's translation.

5 'The City in the Crosshairs: A conversation with Stephen Graham (Pt.1)', http://subtopia.blogspot.com/2007/08/city-in-crosshairs-conversation-with.html.

6 Defined by the US Defense Department as, '[u]nintentional or incidental injury or damage to persons or objects that would not be lawful military targets in the circumstances ruling at the time. Such damage is not unlawful so long as it is not excessive in light of the overall military advantage anticipated from the attack'. US Department of Defense (2001). *Dictionary of Military and Associated Terms*. Defense, US Government.

7 Stephen Graham, '"Clean Territory": Urbicide in the West Bank', www.openDemocracy.net/conflict-politicsverticality/article_241.jsp.

8 United Nations International Criminal Tribunal for the Former Yugoslavia, 'Mladen Naletilic ("Tuta") and Vinko Martinovic ("Stela") Indicted for their Alleged Involvement in the Ethnic Cleansing of the Mostar Municipality', 22 December 1998, The Hague, www.icty.org/sid/7602.

9 According to the Oxford professor of International Relations, Avi Shlaim, 'In the three years after Israel's withdrawal from Gaza, 11 Israelis were killed by rocket fire, prompting the IDF [Israeli Defence Force] to retaliate with military operations that

during the period 2005–7 resulted in the deaths of 1,290 Palestinians in Gaza, including 222 children'. *The Guardian*, 7 January 2009, www.guardian.co.uk/world/2009/jan/07/gaza-israel-palestine.

10 United Nations Human Rights Council, Human Rights in Palestine and Other Occupied Arab Territories: Report of the United Nations Fact Finding Mission on the Gaza Conflict, A/HRC/12/48, 15 September 2009, 10–11.

11 'Endorsing Gaza War Report, UN Human Rights Council Condemns Israel', UN News Center, www.un.org/apps/news/story.asp?NewsID=32578&Cr=palestine &Cr1.

Because the IDF either refused permission or gave only very limited and supervised access to foreign journalists, the treatment of Gazan non-combatants on the ground went almost unreported by the western press. Nevertheless, the UN Mission found evidence of 'continuous and systematic abuse, outrages on personal dignity, humiliating and degrading treatment contrary to fundamental principles of international humanitarian law and human rights law'. UNHRC, ibid. 20. Despite the Goldstone report's findings that both Hamas and Israeli forces were involved in potential war crimes and crimes against humanity during the three-week conflict, the United States described the report as having an 'unbalanced focus', expressing a concern that it will 'exacerbate polarization and divisiveness'. The Israeli Foreign ministry condemned the report for ignoring 'the fact that the Israel Defense Forces took unprecedented measures to avoid harming innocent civilians, and the fact that terror organizations used civilians as human shields in Gaza'. *Haaretz,* 'Delegitimization of Israel Must Be Delegitimized', 17 October 2009, www.haaretz.com/hasen/spages/1121614.html.

12 Avi Shlaim, 'How Israel Brought Gaza to the Brink of Humanitarian Catastrophe', *The Guardian*, 7 January 2009, www.guardian.co.uk/world/2009/jan/07/gaza-israel-palestine.

13 US State Department, *Colombia – Country Reports on Human Rights Practices, 2005*. Bureau of Democracy, Human Rights, and Labor, 8 March 2006.

14 'Bogota's Success Story', Comunidad segura, www.comunidadsegura.org, 8 December 2006.

15 In an operation codenamed 'Sicilian Vespers' the Italian army deployed 150,000 soldiers between 1992 and 1998 in an attempt to show that the government was still in a position to maintain public order after the Mafia assassinations of the investigative judges Giuseppe Falcone and Paolo Borsellino. 'Quando l'esercito sbarca in Sicilia', *Corriere della Sera*, 10 October 2006.

16 Based in the United States Army War College in Pennsylvania.

17 This trend is further examined in the context of William Mulholland's rise and fall during the Los Angeles' 'water wars' in Chapter 9.

18 *Chicago Defender*, 26 November 1955.

19 www.thechicagourbanleague.org.

20 'Chicago Urban League Shifts Focus: Less emphasis on social services, more on economic development', Cheryl Jackson, *News & Views*, 1 April 2007, The New Coalition for Economic and Social Change, Chicago, IL.

21 Although it should be noted that the Zapatista movement consolidated its support most effectively by setting up local Juntas de Buen Gobierno (Councils of Good

Government) in five areas of Chiapas, while contrary to political opportunity structure theory in developed societies 'local and national electoral openings and specific openings to the movement . . . helped to decrease protest activity' (Inclan 2008).

22 Presidential Executive Order No. 661, 12 March 1981, 'Creating a Task Force on the Accelerated Bliss Development of the Tondo Foreshore', www.lawphil.net/executive/execord/eo1981/eo_661_1981.html.

23 http://www.empowerweb.org/newsevents/zoto-rebuilding-lives-manila.

24 Although Brazil's Gini coefficient index declined from 60.7 in 1990 to 57 in 2004, this compared to an index of 45.4 for China in 2002 (where inequality had more than doubled since 1985), and 39.4 for the USA in 2000, which had increased by more than 5 points since 1979. Sweden, one of the least unequal countries as measured by wealth distribution, had a Gini index of 25 in 2005.

25 CIDADE, www.internationalbudget.org,

26 CIDADE, 'For a Popular Sovereignty Network', www.ongcidade.org/site/php/seminario/texto.php?lingua=ingles.

27 An NDPB is defined by the British Cabinet Office as a body which has a role in the processes of national government, but is not a government department or part of one. There are currently over 1,000 such NDPMs in the UK.

28 Greater London Authority, Data Management and Analysis Group, *GLA Demography Update*, 16–2008, October 2008.

29 '"Red Ken" Livingstone Garners Praise from London Capitalists', Bloomberg.com, 19 February 2008.

30 Although winning an Olympic bid does not necessarily make victory in a city election any more likely, as London's twice-elected mayor Ken Livingstone found to his cost in 2008.

31 UNESCO, 'What are Creative Clusters?', http://portal.unesco.org/culture/en/ev.php-URL_ID=29032&URL_DO=DO_TOPIC&URL_SECTION=201.html.

32 Ibid.

33 Richard Florida, 'Our Cities are Good, But They'll Need to be a Lot Better', 11 April 2009, www.creativeclass.com/creative_class/2009/04/16/whos-your-canadian-city-2/.

34 Demos, 'Manchester is Favourite with "New Bohemians"', www.demos.co.uk/press_releases/bohobritain. In 2003 this British left-leaning think-tank used a reduced version of Florida's creativity index to rank the 40 largest British cities according to the size of the gay population, the number of patents issued per head and the degree of ethnic diversity.

35 'When Town Halls Turn to Mecca', *The Economist*, 4 December 2008.

36 Orange, the mobile and broadband arm of France Telecom, launched a £2.6 million multi-media advertising campaign in the UK in September 2008 for its RockCorps volunteering initiative. Its strap-line 'I am who I am because of everyone' appeared in a video filmed in the Heygate public housing estate in Elephant and Castle in London which featured four young girls and their intensive use of mobile phones as part of a community clean-up drive to win tickets for a rap concert.

37 These included the cities of Bradford (where the violence was classified as a riot), Burnley, Oldham, Leeds and Stoke-on-Trent.

38 Breandán Clarke, 'Do Good Walls Make Good Neighbours? Interface walls and the Good Friday Agreement', North Belfast Interface Network, www.nbin.info
39 Ibid.
40 'Nato Defends TV Bombing', http://news.bbc.co.uk/1/hi/world/europe/326653.stm. As Chapter 3 noted, the willingness on the part of military commanders (regardless of the democratic credentials of their regime) to abandon the limitation of a civilian target to that of a war ministry, and to consider facilities such as telecommunications, power-generating stations or bridges to be 'dual use' and therefore targetable, reflects both the importance of communication and information infrastructure as an asset, the loss of which would limit or compromise a society's ability to function, and the enduring importance of the control of the means of communication for both authoritarian and liberal capitalist urban societies.
41 'Concentration of Media Ownership', www.nationmaster.com/encyclopedia/Concentration-of-media-ownership#Europe.
42 'What Happens When a Town Loses Its Newspaper?' *Time Magazine*, 22 March 2009, www.time.com/time/nation/article/0,8599,1886826,00.html.
43 Nielsen Online.
44 Reuters, New York, 15 April, PRNewswire.
45 See for example *The Nation*, 15 June 2009, Editorial; *The Washington Times*, 16 June 2009; *Business Week*, 17 June 2009; *Der Spiegel*, 18 June 2009.
46 'Urbanism', according to Debord, 'is the modern accomplishment of the uninterrupted task which safeguards class power' (Ley 1996: 222–3).
47 *Chinatown* was the first part of a trilogy by the author Robert Towne based on the City of Los Angeles government – the second and third parts were to involve the natural gas and highways departments respectively. Only the second, *The Two Jakes*, was made as a movie sequel to *Chinatown*, directed by and starring Jack Nicholson, which was released in 1990. The final part of the trilogy, *Cloverleaf*, was never made.
48 John Schwartz, 'Louisiana State Fires Hurricane Expert Who Warned of Katrina Flooding', *The Lede*, New York Times Blog, 17 April 2009.
49 Eli Kintisch, 'Hurricane Katrina: Levees Came Up Short, Researchers Tell Congress', *Science*, 11 November 2005, 310.5750, 953–5.
50 The fact that a disproportionate number of New Orleans' flood victims were Latino or Hispanic added more than a little piquancy to Mayor Nagin's self-addressed question 'How Do I Ensure that New Orleans is Not Overrun by Mexican workers?', 'The New Latin Quarter', *Wall Street Journal*, 27 August 2007.
51 For two examples, Senator Rick Santorum (R-PA), 'I mean, you have people who don't heed those warnings and then put people at risk as a result of not heeding those warnings. There may be a need to look at tougher penalties on those who decide to ride it out and understand that there are consequences to not leaving', 6 September 2005; Senator Richard Baker (R-LA), 'We finally cleaned up public housing in New Orleans. We couldn't do it, but God did', *Wall Street Journal*, 12 September 2005.
52 Advancement Project, 'Federal Agencies Sued for Violations of the Fair Housing Act Against New Orleans Public Housing Residents', www.advancementproject.org/.

53 Compossibility is a concept developed by the philosopher Leibniz in order to explain how other contradictory worlds or universes can be simultaneously present in the mind of God. For our purposes, compossibility is used in the sense intended by the contemporary French philosopher Alain Badiou as 'the space of compossibility' where heterogeneous and contradictory truths can co-exist – and not only in the mind of God.

Bibliography

Abers, R. (2000) *Inventing Local Democracy: Grassroots politics in Brazil*. Lynne Rienner Publishers, Boulder, CO.

Agamben, G. (1998) *Homo Sacer: Sovereign power and bare life*. Stanford University Press, Stanford, CA.

—— (2005) *State of Exception*. University of Chicago Press, Chicago, IL and London.

Alexander, C., S. Ishikawa, M. Silverstein et al. (1977) *A Pattern Language: Towns, buildings, construction*. Oxford University Press, New York.

Altheide, D.L. (2002) *Creating Fear: News and the construction of crisis*. Aldine de Gruyter, New York.

Amin, A. (ed.) (1990) *Post-Fordism: A reader*. Blackwell, Oxford.

Anderson, P. (1974) *Passages from Antiquity to Feudalism*. New Left Books, London.

Angeli, L. (2001) 'L'istituto podestarile. Il caso di Torino in prospettiva comparata (1926–45)'. *Passato e presente* 52, 19–40.

Appadurai, A. (1996) *Modernity at Large: Cultural dimensions of globalization*. University of Minnesota Press, Minneapolis and London.

—— (2001) *Globalization*. Duke University Press, Durham, NC and London.

—— (2002) *Transurbanism*. V2-Publishing/NAI Publishers, Rotterdam.

Arsenault, A. and M. Castells (2008) 'The Structure and Dynamics of Global Multi-Media Business Networks'. *International Journal of Communication* 2, 707–48.

Atkinson, R.D., D.K. Correa and J.A. Hedlund (2008) *Explaining International Broadband Leadership*. Information Technology and Innovation Foundation, Washington, DC.

Aunger, R. (2000) *Darwinizing Culture: The status of memetics as a science*. Oxford University Press, Oxford and New York.

Bachrach, P. and M.S. Baratz (1962) 'Two Faces of Power'. *American Political Science Review* 56, 947–52.

Bäck, H., H. Heinelt and A. Magnier (eds) (2006) *The European Mayor: Political leaders in the changing context of local democracy*. VS Verlag für Sozialwissenschaften, Wiesbaden.

Badiou, A. (1999) *Manifesto for Philosophy: Followed by two essays: 'The (re)turn of philosophy itself' and 'Definition of philosophy'*. State University of New York Press, Albany, New York.

Bagnasco, A. and P. Le Galès (eds) (2000) *Cities in Contemporary Europe*. Cambridge University Press, New York.

Baker, C. (2002) *Media, Markets, and Democracy*. Cambridge University Press, Cambridge.

Baldassare, M. and W. Protash (1982) 'Growth Controls, Population Growth, and Community Satisfaction'. *American Sociological Review* 47, 339–46.

Balter, M. (1998) 'The First Cities: Why Settle Down? The Mystery of Communities'. *Science* 282, 1442.

Banfield, E.C. (1961) *Political Influence*. Free Press, Glencoe, IL.

Bang, H.P. (2003) *Governance as Social and Political Communication*. Manchester University Press/Palgrave, Manchester and New York.

Barry, J. (1997) *Rising Tide: The great Mississippi flood of 1927 and how it changed America*. Simon & Schuster, New York.

Bauman, Z. (1999) *In Search of Politics*. Polity, Oxford.

Beauregard, R. 'Theory', paper prepared for a special issue of *Urban Geography* on 'New Directions in Urban Theory' edited by S. Parker and W. Sites (forthcoming).

Benjamin, W. (1973) *Charles Baudelaire: A lyric poet in the era of high capitalism*. New Left Books, London.

—— (1979) *One-way Street and Other Writings*. New Left Books, London.

Bennett, W. (2004) 'Global Media and Politics: Transnational communication regimes and civic cultures'. *Annual Review of Political Science* 7, 125–48.

Berger, S. (ed.) (1981) *Organizing Interests in Western Europe: Pluralism, corporatism and the transformation of politics*. Cambridge University Press, Cambridge.

Berman, M. (1982) *All that is Solid Melts into Air*. Verso, London.

Bhatti, A. (2006) 'Cultural Homogenisation, Places of Memory, and the Loss of Secular Urban Space', in H. Berking, S. Frank, L. Frers, M. Löw, L Meier, S. Steets and S. Stoetzer (eds), *Negotiating Urban Conflicts: Interaction, space and control*, Transcript, Bielefeld.

Binde, P. (1999) 'Nature versus City: Landscapes of Italian Fascism'. *Environment and Planning D: Society and Space* 17, 761–75.

Black, A. (1984) *Guilds and Civil Society in European Political Thought from the Twelfth Century to the Present*. Methuen, London.

Blockmans, W.P. and C. Tilly (1994) *Cities and the Rise of States in Europe, A.D. 1000 to 1800*. Westview Press, Boulder, CO.

Blokland, T. and M. Savage (2001) 'Networks, Class and Place'. *International Journal of Urban and Regional Research* 25, 221–6.

Bollens, S.A. (2008) 'The City, Substate Nationalism, and European Governance'. *Nationalism and Ethnic Politics* 14, 189–222.

Borraz, O. and P. John (2004) 'The Transformation of Urban Political Leadership in Western Europe'. *International Journal of Urban and Regional Research* 28, 107–20.

Boudreau, J.-A. (2005) 'Toronto's Reformist Regime, Municipal Amalgamation and Participatory Democracy', in P. Booth and B. Jouve (eds), *Metropolitan Democracies: Transformations of the state and urban policy in Canada, France and Great Britain*. Ashgate, Aldershot, 99–115.

Boudreau, J.-A, R. Keil and D. Young (2009) *Changing Toronto: Governing urban neoliberalism*. University of Toronto Press, Toronto.

Bourdieu, P. (1986) *Distinction*. Routledge, London and New York.

—— (2000) *Habitus*. Ashgate, Aldershot.

—— (2005) *The Social Structures of the Economy*. Polity, Cambridge and Malden, MA.

Bourdieu, P. and J.B. Thompson (1991) *Language and Symbolic Power*. Harvard University Press, Cambridge, MA.

Bourdieu, P., L.J.D. Wacquant and S. Farage (1994) 'Rethinking the State: Genesis and structure of the bureaucratic field'. *Sociological Theory* 12, 1–18.

Bowker, G. and S. Star (2000) *Sorting Things Out: Classification and its consequences*. The MIT Press, Cambridge, MA and London.

Bowyer, K. (2004) 'Face Recognition Technology: Security versus privacy'. *IEEE Technology and Society Magazine* 23, 9–19.

Braun, H. (2003) *The Assassination of Gaitán: Public life and urban violence in Colombia*. University of Wisconsin Press, Madison.

Brenner, N. (1998) 'Global Cities, Glocal States: Global City Formation and State Territorial Restructuring in Contemporary Europe'. *Review of International Political Economy* 5, 1–37.

—— (2002) 'Decoding the Newest "Metropolitan Regionalism" in the USA: A critical overview'. *Cities* 19, 3–21.

—— (2004) *New State Spaces: Urban governance and the rescaling of statehood*. Oxford University Press, Oxford.

Brenner, N., B. Jessop, M. Jones and G. Macleod (eds) (2003) *State/Space: A reader*. Blackwell, Malden, MA and Oxford.

Brown, J.M.M.G. (2007) 'To Bomb or Not To Bomb? Counterinsurgency, airpower, and dynamic targeting'. *Air and Space Power Journal* 21, 75–85.

Brunet, R. (2002) 'Lignes de forces de l'espace européens'. *Mappemonde* 66.2, 14–19.

Buck, N. and M. Edwards et al. (2002) *Working Capital: Life and Labour in contemporary London*. Routledge, London and New York.

Bullard, R.D. and B. Wright (eds) (2009) *Race, Place, and Environmental Justice after Hurricane Katrina: Struggles to reclaim, rebuild, and revitalize New Orleans and the Gulf Coast*. Westview Press, Boulder, CO.

Burdzhalov, E.N. (2001) 'Russia's Second Revolution: The February 1917 uprising in Petrograd', in M.A. Miller (ed.), *The Russian Revolution: The essential readings*. Blackwell, Malden, MA and Oxford, 39–72.

Burns, P. and M. Thomas (2006) 'The Failure of the Nonregime: How Katrina exposed New Orleans as a regimeless city'. *Urban Affairs Review* 41, 517.

Burrows, R. and D. Beer (2010) 'Rethinking Space: Urban informatics and the sociological imagination', in N. Prior and K. Orton-Johnson (eds), *Rethinking Sociology in the Digital Age*, Palgrave, Basingstoke.

Burrows, R. and N. Gane (2006) 'Geodemographics, Software and Class'. *Sociology* 40, 793–812.

Butler, T. (2007) 'Re-urbanizing London Docklands: Gentrification, suburbanization or new urbanism?' *International Journal of Urban and Regional Research* 31, 759–81.

Butt, G. (1995) *Life at the Crossroads: A history of Gaza*. Rimal Publications, Nicosia.

Calhoun, C. (1998) 'Community without Propinquity Revisited: Communications technology and the transformation of the urban public sphere'. *Sociological Inquiry* 68, 373–97.

Callon, M. (1987) 'Society in the Making: The study of technology as a tool for sociological analysis', in W. Bijker (ed.), *The Social Construction of Technological Systems: New directions in the sociology and history of technology*. MIT Press, London, 83–103.

Cantle, T. (2001) *Community Cohesion: A report of the independent review team*. Home Office, London.

Castells, M. (1977) *The Urban Question*. Edward Arnold, London.

—— (1978) *City, Class and Power*. Macmillan, London.

—— (1983) *The City and the Grassroots*. Edward Arnold, London.

—— (1985) 'From the Urban Question to the City and the Grassroots'. *University of Sussex Urban and Regional Studies Working Paper* No. 47. University of Sussex, Brighton.

—— (1989) *The Informational City: Information technology, economic restructuring, and the urban regional process*. Blackwell, Oxford.

—— (1997) *The Power of Identity*. Blackwell, Oxford.

—— (2000) *The Rise of the Network Society*. Blackwell, Oxford and Malden, MA.

Centre on Housing Rights and Evictions (2007) *Mega Events and Housing Rights*. COHRE, Geneva.

Chan, K.W. (1994) *Cities with Invisible Walls: Reinterpreting urbanization in post-1949 China*. Oxford University Press, Hong Kong and Oxford.

Childe, V.G., T.C. Patterson and C.E. Orser (2004) *Foundations of Social Archaeology: Selected writings of V. Gordon Childe*. AltaMira Press, Walnut Creek, CA.

Clavel, P. (1986) *The Progressive City: Planning and participation, 1969–1984*. Rutgers University Press, New Brunswick, NJ.

Coene, G. and C. Longman (2008) 'Gendering the Diversification of Diversity: The Belgian hijab (in) question'. *Ethnicities* 8, 302–21.

Coleman, R. (2005) 'Surveillance in the City: Primary definition and urban spatial order'. *Crime, Media, Culture* 1, 131.

Consiglio, A. and L. Musella (2005) *La camorra a Napoli*. Guida, Napoli.

Coward, M. (2004) 'Urbicide in Bosnia', in S. Graham (ed.), *Cities, War, and Terrorism: Towards an urban geopolitics*, Blackwell, Malden, MA and Oxford, 154–71.

Cornelius, W.A. (1999) 'Subnational Politics and Democratization: Tensions between center and periphery in the Mexican political system', in W.A. Cornelius, T.A. Eisenstadt and J. Hindley (eds), *Subnational Politics and Democratization in Mexico*. University of California, San Diego, Center for US–Mexican Studies, La Jolla, 3–16.

Cox, K.R. (1993) 'The Local and the Global in the New Urban Politics: A critical review'. *Environment and Planning D* 11, 433–48.

Curran, J. and J. Seaton (2003) *Power without Responsibility: The press, broadcasting, and new media in Britain*. Routledge, London and New York.

Curtis, B. (2001) *The Politics of Population: State formation, statistics, and the census of Canada, 1840–1875*. University of Toronto Press.

Cutter, S. and C. Emrich (2006) 'Moral Hazard, Social Catastrophe: The changing face of vulnerability along the hurricane coasts'. *The Annals of the American Academy of Political and Social Science* 604, 102.

Dahl, R. (1961) *Who Governs? Democracy and power in an American city*. Yale University Press, New Haven, CT and London.

Daniel, G. (1968) *The First Civilisations: The archaeology of their origins*. Thames & Hudson, London.

Davies, J. and D.L. Imbroscio (eds) (2009) *Theories of Urban Politics*. Sage, London and Thousand Oaks, CA.

Davies, P.J. (1997) 'The Politics of Perpetuation: Trajan's Column and the art of commemoration'. *American Journal of Archaeology* 101, 41–65.

Davis, M. (1992) *City of Quartz: Excavating the future of Los Angeles*. Vintage, London.

—— (2006) *Planet of Slums*. Verso, London.

Deas, I. and K.G. Ward (2000) 'From the "New Localism" to the "New Regionalism"? The implications of regional development agencies for city-regional relations'. *Political Geography* 19, 273–92.

De Certeau, M. (1984) *The Practice of Everyday Life*. University of California Press, Berkeley.

DeFilippis, J. (1999) 'Alternatives to the "New Urban Politics": Finding locality and autonomy in local economic development'. *Political Geography* 18, 973–90.

Degler, C.N. (1968) 'Political Parties and the Rise of the City', in L.W. Dorsett (ed.), *The Challenge of the City 1860–1910*, D.C. Heath & Co., Lexington, MA, 100–14.

De Krey, G.S. (1985) *A Fractured Society: The politics of London in the first age of party 1688–1715*. Clarendon, Oxford.

Deleuze, G. (1976) *Rhizomé: Introduction*. Editions De Minuit, Paris.

Deleuze, G. and F. Guattari (1983a) *Anti-Oedipus: Capitalism and schizophrenia*. Athlone, 1984, London.

Deleuze, G. and F. Guattari (1983b) *On the Line*. Semiotext(e), New York.

Demoen, K. (2001) *The Greek City from Antiquity to Present: Historical reality, ideological construction, literary representation*. Peeters, Louvain; Sterling, VA.

De Ste Croix, G.E.M. (1989) *The Class Struggle in the Ancient Greek World: From the archaic age to the Arab conquests*. Cornell University Press, Ithaca, NY.

Diamond, J.M. (2005a) *Collapse: How societies choose to fail or succeed*. Viking, New York.

—— (2005b) *Guns, Germs, and Steel: The fates of human societies*. W.W. Norton, New York.

Diamond, J.M. and T.J. Case (1986) *Community Ecology*. Harper & Row, New York.

DiGaetano, A. and J.S. Klemanski (1999) *Power and City Governance: Comparative perspectives on urban development*. University of Minnesota Press, Minneapolis.

Dodge, M. and R. Kitchin (2004) 'Flying through Code/Space: The real virtuality of air travel'. *Environment and Planning A* 36, 195–212.

—— (2010) *Code/Space: Software and everyday life*. MIT Press, Cambridge, MA.

Dreier, P., J. Mollenkopf and T. Swanstrom (2001) *Place Matters. Metropolitics for the twenty-first century*. University Press of Kansas, Lawrence, KA.

Durkheim, E. (1984) *The Division of Labour in Society*. Macmillan, Basingstoke.

Durliat, J. (1988) 'Le salaire de la paix sociale dans les royaumes barbares (Ve–Vie siècles)', in H. Wolfram and A. Schwarcz (eds), *Anerkennung und Integration: Zu den wirtschaftlichen Grundlagen der Völkerwanderungszeit, 400–600*. VöAW, Vienna, 21–72.

Dyson, M.E. (2006) *Come Hell or High Water: Hurricane Katrina and the color of disaster*. Basic Civitas, New York.

Elkin, S. (1987) *City and Regime in the American Republic*. University of Chicago Press, Chicago, IL.

Ellin, N. (1999) *Postmodern Urbanism*. Princeton Architectural Press, New Jersey.

Epstein, S.R. (2001) 'Introduction: Town and country in Europe, 1300–1800'. *Themes in International Urban History 5*, Cambridge University Press, Cambridge, 1–29.

Esposito, R. (2008) *Bíos: Biopolitics and philosophy*. University of Minnesota Press, Bristol and Minneapolis, MI.

Ferman, B. (1996) *Challenging the Growth Machine: Neighborhood politics in Chicago and Pittsburgh*. University Press of Kansas, Lawrence.

Fife, G. (2004) *The Terror: The shadow of the guillotine. France, 1792–1794*. Portrait, London.

Finley, M.I. (1983) *Politics in the Ancient World*. Cambridge University Press, Cambridge and New York.

Fisk, R. (2005) *The Great War for Civilisation: The conquest of the Middle East*. Alfred Knopf, New York.

Florida, R. (2003) 'Cities and the Creative Class'. *City & Community* 2, 3–19.

—— (2005) *Cities and the Creative Class*. Routledge, London.

—— (2008) *Who's Your City?: How the creative economy is making where to live the most important decision of your life*. Basic Books, New York.

Flusty, S. (2004) *De-Coca-colonization: Making the globe from the inside out*. Routledge, New York and London.

—— (2005) 'Postmodernism', in D. Atkinson (ed.) *Cultural Geography: A critical dictionary of key ideas*. I.B. Tauris, London, 169–74.

Foucault, M. (1980) *Power/Knowledge: Selected interviews and other writings, 1972–1977*. Harvester Press, Brighton.

—— (1982) *I, Pierre Rivière, having slaughtered my mother, my sister, and my brother — a case of parricide in the 19th century*. University of Nebraska Press, Lincoln, NE.

Foucault, M., G. Burchell, C. Gordon and P. Miller (1991) *The Foucault Effect: Studies in governmentality with two lectures by and an interview with Michel Foucault*. University of Chicago Press, Chicago, IL.

Friedmann, J. (2005) *China's Urban Transition*. University of Minnesota Press, Minneapolis, MI and Bristol.

Friedmann, J. and G. Wolff (1982) 'World City Formation: An agenda for research and action'. *International Journal of Urban and Regional Research* 6, 309–43.

Friedmann, J. and R.M. Wulff (1976) *The Urban Transition: Comparative studies of newly industrializing societies*. Edward Arnold, London.

Friedrichs, C.R. (2000) *Urban Politics in Early Modern Europe*. Routledge, London and New York.

Fujita, K. and R. Child Hill (1995) 'Global Toyotaism and Local Development'. *International Journal of Urban and Regional Research* 19, 7–22.

Gandy, M. (1999) 'The Paris Sewers and the Rationalization of Urban Space'. *Transactions of the Institute of British Geographers* NS 24, 23–44.

Giddens, A. (1985) *A Contemporary Critique of Historical Materialism*. Polity, London.

—— (1993) *New Rules of Sociological Method: A positive critique of interpretative sociologies*. Polity, Oxford.

—— (2003) *Runaway World: How globalisation is reshaping our lives*. Routledge, New York.

Ginsborg, P. (2005) *Silvio Berlusconi: Television, power and patrimony*. Verso, London.

Giraldo, J. (1996) *Colombia: The genocidal democracy*. Common Courage, Monroe, ME.

Giugni, M. (1999) 'How Social Movements Matter: Past research, present problems, future developments', in M. Giugni, D. McAdam and C. Tilly (eds), *How Social Movements Matter*, Minnesota University Press, Minneapolis and London, xiii–xxxiii.

Glasberg, D. (1988) 'The Political Economic Power of Finance Capital and Urban Fiscal Crisis: Cleveland's Default, 1978'. *Journal of Urban Affairs* 10, 219–39.

Gleeson, B. (1998) 'Justice and the Disabling City', in R. Fincher and J.M. Jacobs (eds), *Cities of Difference*, Guilford Press, New York and London, 89–119.

Glick, E.F. (2003) 'Harlem's Queer Dandy: African American modernism and the artifice of blackness'. *Modern Fiction Studies* 49, 411–42.

Goffart, W. (2006) 'The Barbarians in Late Antiquity and How They Were Accommodated in the West' in T.F.X. Noble (ed.), *From Roman Provinces to Medieval Kingdoms*. Routledge, London, 235–61.

Goffredo, M., J. Carter and M. Nixon (2008) *Front-view Gait Recognition*. University of Southampton, Southampton.

Goldhagen, D. (1996) *Hitler's Willing Executioners*. Alfred Knopf, New York.

Goldsmith, M. (2001) 'Urban Governance', in R. Paddison (ed.), *Handbook of Urban Studies*, Sage, London, 325–35.

Goldsmith, M. and H. Larsen (2004) 'Local Political Leadership: Nordic Style'. *International Journal of Urban and Regional Research* 28, 121–33.

Goodman, R. (1979) *The Last Entrepreneurs: America's regional wars for jobs and dollars*. Simon & Schuster, New York.

Goodwin, M. and J. Painter (1996) 'Local Governance, the Crises of Fordism and the Changing Geographies of Regulation'. *Transactions of the Institute of British Geographers* 21, 635–48.

Goss, J. (1995) ' "We Know Who You Are and We Know Where You Live": The instrumental rationality of geodemographic systems'. *Economic Geography* 71, 171–98.

Government Office for London (2006) *Government Office for London Explained*. HMSO, London.

Graham, S. (ed.) (2004) *Cities, War and Terrorism: Towards an urban geopolitics*. Blackwell, Malden, MA and Oxford.

—— (2010) *Cities Under Siege: The new military urbanism*. Verso, London.

Graham, S. and S. Marvin (2001) *Splintering Urbanism: Technology, globalization and the networked metropolis*. Routledge, London and New York.

Graham, S. and D. Wood (2007) 'Digitizing Surveillance: Categorization, space, inequality', in S.P. Hier and J. Greenberg (eds), *The Surveillance Studies Reader*, Open University/McGraw-Hill, Maidenhead and New York, 218–30.

Gramsci, A. (1971) 'The Modern Prince', in *Selections from the Prison Notebooks*, Lawrence & Wishart, London, 125–205.

Granovetter, M.S. (1973) 'The Strength of Weak Ties'. *American Journal of Sociology* 78, 1360–80.

Greer, I. (2007) 'Special Interests and Public Goods. Organized labor's coalition politics in Hamburg and Seattle', in L. Turner, D. Cornfield, P. Evans (eds) *Labor in the New Urban Battlegrounds: Local solidarity in a global economy*, Cornell University Press, Ithaca, 193–210.

Gregory, D. and J. Urry (1985) *Social Relations and Spatial Structures*. Cambridge University Press, Cambridge.

Gret, M. and Y. Sintomer (2005) *The Porto Alegre Experiment: Learning lessons for better democracy*. Zed Books, London and New York.

Griffin, M. and F.A Kittler (1996) 'The City is a Medium'. *New Literary History* 24, 717–29.

Grodach, C. (2002) 'Reconstituting Identity and History in Post-war Mostar, Bosnia-Herzegovina'. *City* 6, 61–82.

Grusin, R.A. (2004) 'Premediation'. *Criticism* 46, 17–39.

Haber, P. (2006) *Power from Experience: Urban popular movements in late twentieth-century Mexico*. Pennsylvania State University Press, University Park.

Habermas, J. (1989) *The Structural Transformation of the Public Sphere*. MIT Press, Cambridge, MA.

Hackworth, J.R. (2007) *The Neoliberal City: Governance, ideology, and development in American urbanism*. Cornell University Press, Ithaca.

Haggerty, K. (2006) 'Tear Down the Walls: On demolishing the Panopticon', in D. Lyon (ed.), *Theorizing Surveillance: The Panopticon and beyond*, Willan Publishing, Cullompton, 23–45.

Hall, M. (2006) 'Urban Entrepreneurship, Corporate Interests and Sports Mega-events: The thin policies of competitiveness with the hard outcomes of neoliberalism'. *Sociological Review* 54, 59–70.

Hall, P. (1998) *Cities in Civilisation: Culture, innovation and urban order*. Weidenfeld & Nicolson, London.

Hall, T. and P. Hubbard (1996) 'The Entrepreneurial City: New urban politics, new urban geographies?' *Progress in Human Geography* 20, 153–74.

Hamel, P., M. Mayer and H. Lustiger-Thaler (eds) (2000) *Urban Movements in a Globalising World*. Routledge, London.

Hannerz, U. (1980) *Exploring the City*. Columbia University Press, New York.

Harding, A. (1994) 'Urban Regimes and Growth Machines: Towards a cross-national research agenda'. *Urban Affairs Quarterly* 29, 356–82.

—— (2009) 'The History of Community Power', in J.S. Davies and D.L. Imbroscio (eds), *Theories of Urban Politics*. 2nd edn, Sage, London, 27–39.

Hartman, C.W. and G.D. Squires (2006) *There Is No Such Thing as a Natural Disaster: Race, class, and Hurricane Katrina*. Routledge, New York and London.

Harvey, D. (1989a) 'From Managerialism to Entrepreneurialism: The transformation in urban governance in late capitalism'. *Geografiska Annaler. Series B, Human Geography* 71, 3–17.

—— (1989b) *The Condition of Postmodernity*. Basil Blackwell, Oxford.

—— (1989c) *The Urban Experience*. Basil Blackwell, Oxford.

—— (2006) 'The Political Economy of Public Space', in S.M. Low and N. Smith (eds), *The Politics of Public Space*, Routledge, New York and Abingdon, 17–34.

Häussermann, H. and A. Haila (2004) 'The European City: A conceptual framework and normative project', in Y. Kazepov (ed.), *Cities of Europe: Changing contexts, local arrangements, and the challenge to urban cohesion*, Blackwell, Oxford, 43–63.

Hennock, E.P. (2007) *The Origin of the Welfare State in England and Germany, 1850–1914: Social policies compared*. Cambridge University Press, Cambridge.

Hirschman, A.O. (1972) *Exit, Voice and Loyalty: Response to decline in firms, organizations and states*. Harvard University Press, Boston.

Hodder, I. (ed.) (1996) *On the Surface: Catalhoyuk 1993–1995*. McDonald Institute of Archaeological Research and the British Institute of Archaeology at Ankara, Cambridge.

—— (2002) 'Ethics and Archaeology: The attempt at Catalhoyuk'. *Near Eastern Archaeology* 65, 174–81.

Howell, M.C. (2000) 'The Spaces of Late Medieval Urbanity', in M. Boone and P. Stabel (eds), *Shaping Urban Identity in Late Medieval Europe*, Garant, Leuven-Apeldoorn, 3–23.

Hunter, F. (1953) *Community Power Structure: A study of decision makers*. University of North Carolina Press, Chapel Hill.

—— (1959) *Top Leadership, U.S.A.* University of North Carolina Press, Chapel Hill.

Imrie, R., L. Lees and M. Raco (2009) *Regenerating London: Governance, sustainability and community in a global city*. Routledge, London and New York.

Inclan, M. (2008). 'From the Ya Basta! to the Caracoles: Zapatista mobilization under transitional conditions.' *American Journal of Sociology* 113, 1316–50.

Jacobs, J. (1985) *Cities and the Wealth of Nations*. Viking, London.

—— (2004) *Dark Age Ahead*. Random House, New York.

Jameson, F. (1971) *Marxism and Form: Twentieth-century dialectical theories of literature*. Princeton University Press, Princeton, NJ.

Jarman, N. (2005) *No Longer A Problem? Sectarian violence in Northern Ireland.* Institute for Conflict Research, Belfast.

—— (2008) *Interface and Security Barrier Mapping Project.* Institute for Conflict Research, Belfast.

Jessop, B. (1992) 'Fordism and Post-Fordism: A critical reformulation', in M. Storper and A.J. Scott (eds), *Pathways to Industrialization and Regional Development*, Routledge, London, 46–69.

—— (2000) 'Globalisation, Entrepreneurial Cities and the Social Economy', in P. Hamel, H. Lustiger-Thaler and M. Mayer (eds), *Urban Movements in a Globalising World*, Routledge, London and New York, 81–100.

—— (2001) 'Regulationist and Autopoieticist Reflections on Polanyi's Account of Market Economies and the Market Society'. *New Political Economy* 6, 213–32.

Jonas, A.E.G. and D. Wilson (1999) 'The City as a Growth Machine: Critical reflections two decades later', in A.E.G. Jonas and D. Wilson (eds), *The Urban Growth Machine: Critical perspectives two decades later*, State University of New York Press, Wantage, 3–18.

Jones, K. (2000) *The Making of Social Policy in Britain: From the Poor Law to New Labour*. Athlone, London.

Judd, D.R. and T. Swanstrom (2008) *City Politics: The political economy of urban America.* Pearson Longman, New York and London.

Judge, D., G. Stoker and H. Wolman (eds) (1995) *Theories of Urban Politics*. Sage, London.

Jusova, I. (2008) 'Hirsi Ali and van Gogh's Submission: Reinforcing the Islam vs. women binary'. *Women's Studies International Forum* 31, 148–55.

Kaika, M. (2005) *City of Flows: Modernity, nature, and the city*. Routledge, New York.

Kaika, M. and E. Swyngedouw (2000) 'Fetishizing the Modern City: The phantasmagoria of urban technological networks'. *International Journal of Urban and Regional Research* 24, 120–38.

Keating, M. (1991) *Comparative Urban Politics. Power and the city in the United States, Canada, Britain and France*. Edward Elgar, Aldershot.

Keil, R. (2003) 'Globalization Makes States: Perspectives on local governance in the age of the world city', in N. Brenner, B. Jessop, M. Jones and G. Macleod (eds), *State/Space: A reader*, Blackwell, Malden, MA and Oxford, 278–95.

Keil, R. and S. Kipfer (2002) 'Toronto Inc? Planning the competitive city in the new Toronto'. *Antipode* 34, 227–64.

Keil, R. and R. Mahon (2009) *Leviathan Undone? Towards a political economy of scale.* University of British Columbia Press, Vancouver, BC and Toronto.

Kennedy, P.T. (2010) *Local Lives and Global Transformations: Towards world society.* Palgrave Macmillan, Basingstoke.

Kerr, T., E. Wood, D. Small, A. Palepu and M. Tyndall (2003) 'Potential Use of Safer Injecting Facilities among Injection Drug Users in Vancouver's Downtown Eastside'. *Canadian Medical Association Journal* 169, 759.

Kirby, A. and T. Abu-Rass (1999) 'Employing the Growth Machine Heuristic in a Different Political and Economic Context: The case of Israel', in A.E.G. Jonas and

D. Wilson (eds), *The Urban Growth Machine: Critical perspectives two decades later*, State University of New York Press,Wantage, 213–25.
Kittler, F. (1996) 'The History of Communication Media', Special Issue C-Theory.net, www.ctheory.net/articles.aspx?id=45.
Kittler, F. and M. Griffin (1996) 'The City is a Medium'. *New Literary History*, 717–29.
Kohl, P. and R. Wright (1977) 'Stateless Cities: The differentiation of societies in the near eastern Neolithic'. *Dialectical Anthropology* 2, 271–83.
Krätke, S. and P. Taylor (2004) 'A World Geography of Global Media Cities'. *European Planning Studies* 12, 459–77.
Kübler, D. and P. Michel (2006) 'Mayors in Vertical Power Relations', in H. Bäck, H. Heinelt and A. Magnier (eds), *The European Mayor: Political leaders in the changing context of local democracy*, VS Verlag für Sozialwissenschaften, Wiesbaden, 221–44.
Lash, S. (2002) *Critique of Information*. Sage, London.
Latour, B. (2005) *Reassembling the Social: An introduction to actor-network theory*. Oxford University Press, Oxford.
Law, J. (1999) 'After ANT: Complexity, naming and topology', in J. Law and J. Hassard (eds) *Actor Network Theory and After*. Blackwell, Oxford, 1–14.
Lefebvre, H. (1991) *The Production of Space*. Blackwell, Oxford.
Leibovitz, J. (1999) 'New Spaces of Governance: Re-reading the local state in Ontario'. *Space and Polity* 3, 199–216.
Leitner, H., J. Peck and E.S. Sheppard (2007) *Contesting Neoliberalism: Urban frontiers*. Guilford Press, New York and London.
Lewis-Williams, D. (2004) 'Constructing a Cosmos: Architecture, power and domestication at Catalhoyuk'. *Journal of Social Archaeology* 4, 28–59.
Ley, D. (1991) 'Gentrification: A ten year overview', in K. Gerecke (ed.), *The Canadian City*, Black Rose Books, Montreal and New York, 181–96.
—— (1996) *The New Middle Class and the Remaking of the Central City*. Oxford University Press, Oxford.
Liebeschütz, W. (1992) 'The End of the Ancient City', in J. Rich (ed.), *The City in Late Antiquity*, Routledge, London and New York, 1–49.
—— (2006) 'Cities, Taxes and the Accommodation of the Barbarians: The theories of Durliat and Goffard', in T.F.X. Noble (ed.), *From Roman Provinces to Medieval Kingdoms*, Routledge, London, 309–23.
Lijphart, A. (2008) *Verzuiling, pacificatie en kentering in de Nederlandse politiek*. Amsterdam University Press, Amsterdam.
Lindblom, C. (1959) 'The Science of "Muddling Through"'. *Public Administration Review* 19, 79–88.
Lipietz, A. (2003) 'The National and the Regional: Their autonomy vis-a-vis the capitalist world crisis', in N. Brenner, B. Jessop, M. Jones and G. Macleod (eds), *State/Space: A reader*, Blackwell, Malden, MA and Oxford, 239–55.
Lipset, S.M. and S. Rokkan (eds) (1967) *Party Systems and Voter Alignments: Cross-national perspectives*. Free Press, New York.

Lipsky, M. (1980) *Street Level Bureaucracy: Dilemmas of the individual in public services*. Russell Sage Foundation, New York.

Livingstone, G. (2003) *Inside Colombia: Drugs, democracy and war*. Latin America Bureau, London.

Logan, J. and H. Molotch (2007) *Urban Fortunes: The political economy of place*. University of California Press, Berkeley.

Logan, J., R.B. Whaley and K. Crowder (1999) 'The Character and Consequences of Growth Regimes: An assessment of twenty years of research', in A.E.G. Jonas and D. Wilson (eds), *The Urban Growth Machine: Critical perspectives two decades later*, State University of New York Press, Wantage, 73–93.

Logan, J.R. (1978) 'Growth, Politics, and the Stratification of Places'. *The American Journal of Sociology* 84, 404–16.

—— (2002) *The New Chinese City: Globalization and market reform*. Blackwell Publishers, Oxford.

Low, M. (2007) 'Political Parties and the City: Some thoughts on the low profile of partisan organisation and mobilisation in urban political theory'. *Environment and Planning A* 39, 2652–67.

Low, S.M. and N. Smith (eds) (2006) *The Politics of Public Space*. Routledge, New York and London.

Lynd, R.S. and H.M. Lynd (1929) *Middletown*. Harcourt Brace, New York.

—— (1937) Middletown in Transition. Harcourt Brace, New York.

Lyon, D. (2007) 'Surveillance, Power and Everyday Life', in R. Mansell (ed.), *The Oxford Handbook of Information and Communication Technologies*, Oxford University Press, Oxford, 449–72.

—— (2008) 'Biometrics, Identification and Surveillance'. *Bioethics* 22, 499–508.

MacLeod, G. (2001) 'New Regionalism Reconsidered: Globalization and the remaking of political economic space'. *International Journal of Urban and Regional Research* 25, 804–29.

Maisels, C.K. (1993) *The Emergence of Civilization: From hunting and gathering to agriculture, cities and the state in the Near East*. Routledge, London and New York.

Mallory, S. (2007) *Understanding Organized Crime*. Jones & Bartlett Publishers, Mississauga, Ontario.

Maschner, H.D.G. (2003) 'Historical Traditions and Darwinian Theory'. *Cambridge Archaeological Journal* 13, 283–5.

Mattox, R.M. and P.S. Rodgers (2007) *Counterinsurgency in the 21st Century: The foundation and implications of the new U.S. doctrine*. Naval Postgraduate School, Monterey.

McDowell, L. (2009) *Working Bodies: Interactive service employment and workplace identities*. Wiley-Blackwell, Oxford.

McLuhan, M. (2001) *Understanding Media*. Routledge, London and New York.

McLuhan, M. and Q. Fiore (1967) *The Medium is the Massage*. Allen Lane, London.

McQuire, S. (2008) *The Media City: Media, architecture and urban space*. Sage, Los Angeles, CA.

Meece, S. (2006) 'A Bird's Eye View of a Leopard's Spots: The Çatalhöyük "map" and the development of cartographic representation in prehistory'. *Anatolian Studies* 56, 1–16.

Meier, W. (2002) 'Media Ownership—Does It Matter?', in R. Mansell, R. Samarajiva and A. Mahan (eds), *Networking Knowledge for Information Societies: Institutions and intervention*, Delft University Press, Delft, 298–302.

Mellaart, J. (1967) *Çatal Hüyük: A neolithic town in Anatolia*. Thames & Hudson, London.

Michels, R. (1966) *Political Parties*. Free Press, New York.

Miner, H. (1952) 'The Folk-Urban Continuum'. *American Sociological Review* 17, 529–37.

Mollenkopf, J. (1992) *A Phoenix in the Ashes: The rise and fall of the Koch coalition in New York City politics*. Princeton University Press, Princeton, NJ.

—— (1995) 'Community Power in a Postreform City: Politics in New York City'. *American Political Science Review* 89, 502–3.

Molotch, H. (1976) 'The City as a Growth Machine: Toward a political economy of place'. *American Journal of Sociology* 82, 309–30.

—— (2009) *City Durables: Encounters, worries, and their devices*. City Institute Guest Lecture, York University, Toronto.

Moore, B. (1966) *Social Origins of Dictatorship and Democracy: Lord and peasant in the making of the modern world*. Allen Lane, London.

Mörner, M. (1993) *Region and State in Latin America's Past*. Johns Hopkins University Press, Baltimore and London.

Morozov, E. (2009) 'Iran: Downside to the "Twitter Revolution"'. *Dissent* 56, 10–14.

Morris, I. (1991) 'The Early Polis as a City State', in J. Rich and A. Wallace-Hadrill (eds), *City and Country in the Ancient World*, Routledge, London and New York, 25–58.

Morse, R.M. (1962) 'Some Characteristics of Latin American Urban History'. *The American Historical Review* 67, 317–38.

Moser, C. and C. McIlwaine (1999) 'Participatory Urban Appraisal and Its Application for Research on Violence'. *Environment & Urbanization* 11, 203–26.

—— (2004) *Encounters with Violence in Latin America: Urban poor perceptions from Colombia and Guatemala*. Routledge, New York and London.

Mossberger, K. And G. Stoker (2001) 'The Evolution of Urban Regime Theory: The challenge of conceptualization'. *Urban Affairs Review* 36, 810–35.

Moynihan, D.P. (1993) 'When the Irish Ran New York'. *City Journal* Spring,www.city-journal.org/article02.php?aid=1499.

Mumford, L. (1964) 'Authoritarian and Democratic Technics'. *Technology and Culture* 5, 1–8.

—— (1989) *The City in History*. Harcourt, Orlando, FL.

Murakami Wood, D. and S. Graham (2006) 'Permeable Boundaries in the Software-sorted Society: Surveillance and differentiations of mobility', in M. Sheller and J. Urry (eds), *Mobile Technologies of the City*, Routledge, London and New York, 177–91.

Narang-Suri, S. (2009) 'Urban Planning and Post-war Reconstruction under Transitional Administrations: The case of Mostar'. Unpublished PhD thesis, Department of Politics, University of York, York.

Nevola, F. (2007) *Siena: Constructing the Renaissance city*. Yale University Press, New Haven, CT and London.

Newsinger, J. (2002) *British Counterinsurgency: From Palestine to Northern Ireland*. Palgrave, Houndmills, Basingstoke, Hampshire and New York.

Nicholls, W.J. (2008) 'The Urban Question Revisited: The importance of cities for social movements'. *International Journal of Urban and Regional Research* 32, 841–59.

Nyanmnjoh, F.B. (2004) 'Media Ownership and Control in Africa in the Age of Globalization', in P. Thomas and Z. Nain (eds), *Who Owns the Media?: Global trends and local resistances,* Zed Books, Penang, Malaysia, 119–34.

Oates, J. (1993) 'Trade and Power in the Fifth and Fourth Millennia BC: New evidence from northern Mesopotamia'. *World Archaeology* 24, 403–22.

O'Connor, J. (1973) *The Fiscal Crisis of the State*. St Martin's Press, New York.

Ofek, H. (2001) *Second Nature: Economic origins of human evolution*. Cambridge University Press, Cambridge.

Olds, K. and H. Yeung (2004) 'Pathways to Global City Formation: A view from the developmental city-state of Singapore'. *Review of International Political Economy* 11, 489–521.

Organisation for Economic Co-operation and Development. (2006) *Competitive Cities in the Global Economy*. OECD, Paris.

—— (2007) *Competitive Cities: A new entrepreneurial paradigm in spatial development*. OECD, Paris.

Osborne, R. (1987) *Classical Landscape with Figures*. Sheridan House, London.

Osborne, T. and N. Rose (1999) 'Governing Cities: Notes on the spatialisation of virtue'. *Environment and Planning D: Society and Space* 17, 737–60.

Pahl, R.E. (1970) *Whose City? and Other Essays on Sociology and Planning*. Longman, London.

Pak, S.Y. (2006) 'Politicizing Imagery and Representation of Muslim Womanhood: Reflections on the Islamic headscarf controversy in Turkey'. *Asian Journal of Women's Studies* 12, 32–60.

Palliser, D.M. (2006) *Towns and Local Communities in Medieval and Early Modern England*. Ashgate, Aldershot.

Park, R.E., E.W. Burgess and R.D. McKenzie (1925) *The City*. University of Chicago Press, Chicago and London.

Parker, S. (2000) 'Tales of the City: Situating urban discourse in place and time'. *City* 4, 233–46.

—— (2001a) 'Community, Social Identity and the Structuration of Power in the Contemporary European City. Part One: Towards a theory of urban structuration'. *City* 5, 189–202.

—— (2001b) 'Community, Social Identity and the Structuration of Power in the Contemporary European City. Part Two: Power and identity in the urban community: A comparative analysis'. *City* 5, 281–309.

—— (2004) *Urban Theory and the Urban Experience: Encountering the city*. Routledge, London and New York.

—— (2006) 'Managing the Political Field: Italian regions and the territorialisation of politics in the Second Republic'. *Journal of Southern Europe and the Balkans* 8, 235–53.

Parker, S., E. Uprichard and R. Burrows (2007) 'Class Places and Place Classes: Geodemographics and the spatialization of class'. *Information Communication and Society* 10, 902–21.

Peck, J. (2005) 'Struggling with the Creative Class'. *International Journal of Urban and Regional Research* 29, 740–70.

Peck, J. and A. Tickell (2002) 'Neoliberalizing Space'. *Antipode* 34, 380–404.

Peters, B.G. (2004) 'Politics is about Governing', in A. Leftwich (ed.), *What is Politics?*, Polity, Cambridge, 23–40.

Peterson, P.E. (1981) *City Limits*. University of Chicago Press, Chicago.

Picard, R. (2002) *The Economics and Financing of Media Companies*. Fordham University, New York.

Pinder, D. (2005) *Visions of the City: Utopianism, power and politics in twentieth-century urbanism*. Edinburgh University Press.

Pine, J. (2008) 'Contact, Complicity, Conspiracy: Affective communities and economies of affect in Naples'. *Law, Culture & the Humanities* 4, 201–23.

Plato (1993) *Republic*. Oxford University Press, Oxford and New York.

Powell, J.A., H.K. Jeffries and E. Stiens (2006) 'Towards a Transformative View of Race: The Crisis and Opportunity of Katrina', in C.W. Hartman and G.D. Squires (eds), *There Is No Such Thing as a Natural Disaster: Race, class, and Hurricane Katrina*, Routledge, New York; London, 59–84.

Putnam, R.D., with R. Nanetti and R. Leonardi (1993) *Making Democracy Work: Civic traditions in modern Italy*. Princeton University Press, Princeton, NJ.

Qiu, J.L. (2009) *Working-class Network Society: Communication technology and the information have-less in urban China*. MIT, Cambridge, MA and London.

Raaflaub, K.A. (2005) *Social Struggles in Archaic Rome: New perspectives on the conflict of the orders*. Blackwell, Oxford.

Rama, A. and J.C. Chasteen (1996) *The Lettered City*. Duke University Press, Durham, NC.

Rex, J., Moore, A. et al. (1967) *Race, Community and Conflict: A study of Sparkbrook*. Institute of Race Relations/Oxford University Press, Oxford.

Robinson, J. (2006) *Ordinary Cities: Between modernity and development*. Routledge, London and New York.

Rothman, M.S. (2004) 'Studying the Development of Complex Society: Mesopotamia in the late fifth and fourth millennia BC'. *Journal of Archaeological Research* 12, 75–119.

Rubin, J.W. (1999) 'Zapotec and Mexican: Ethnicity, militancy, and democratization in Juchitán, Oaxaca', in W.A. Cornelius, T.A. Eisenstadt and J. Hindley (eds), *Subnational Politics and Democratization in Mexico*, Center for US–Mexican Studies, University of California, San Diego, La Jolla, 175–206.

Rubinstein, N. (1958) 'Political Ideas in Sienese Art: The frescoes by Ambrogio Lorenzetti and Taddeo di Bartolo in the Palazzo Pubblico'. *Journal of the Warburg and Courtauld Institutes* 21, 179–207.

Said, E. (2003) *Orientalism*. Penguin, London.

—— (2004) 'Orientalism Once More'. *Development & Change* 35, 869.

Sanín, F.G. and A.M. Jaramillo (2004) 'Crime, (Counter-)insurgency and the Privatization of Security: The case of Medellín, Colombia'. *Environment and Urbanization* 16, 17–30.

Sassen, S. (2000) *Cities in a World Economy*. Pine Forge Press, Thousand Oaks, CA.

—— (2001) *The Global City: New York, London, Tokyo*. Princeton University Press, Princeton, NJ and Oxford.

—— (2008) 'Cities and New Wars: After Mumbai', OpenDemocracy, www.opendemocracy.net/article/the-new-wars-and-cities-after-mumbai-0.

Saunders, P. (1986) *Social Theory and the Urban Question*. Routledge, London.

Savage, M., G. Bagnall and B. Longhurst (2004) *Globalization and Belonging*. Sage, London and Thousand Oaks, CA.

Savage, M. and R. Burrows (2007) 'The Coming Crisis of Empirical Sociology'. *Sociology* 41.

—— (2009) 'Some Further Reflections on the Coming Crisis of Empirical Sociology'. *Sociology* 43, 762–72.

Saviano, R. (2007) *Gomorrah*. Macmillan, London.

Schiesl, M.J. (1977) 'Politicians in Disguise: The changing role of public administrators in Los Angeles, 1900–1920', in M.H. Ebner and E.M. Tobin (eds), *The Age of Urban Reform: New perspectives on the Progressive Era*, Kennikat Press, Port Washington and London, 102–16.

Schneider, J. and I. Susser (eds) (2004) *Wounded Cities: Destruction and reconstruction in a globalized world*. Berg, Oxford.

Schneider, M. and P. Teske (1993) 'The Antigrowth Entrepreneur: Challenging the "equilibrium" of the growth machine'. *The Journal of Politics* 55, 720–36.

Schneider, S.K. (2005) 'Administrative Breakdowns in the Governmental Response to Hurricane Katrina'. *Public Administration Review* 65, 515–16.

Schniedewind, W.M. (2004) *How the Bible Became a Book: The textualization of ancient Israel*. Cambridge University Press, Cambridge.

Schumpeter, J.A. (1987) *Capitalism, Socialism and Democracy*. Unwin Paperbacks, London.

Schuurman, F.J. and T.v. Naerssen (eds) (1988) *Urban Social Movements in the Third World*. Routledge, London.

Sennett, R. (1970) *The Uses of Disorder: Personal identity and city life*. Alfred Knopf, New York.

—— (1990) *The Conscience of the Eye: The design and social life of cities*. Faber & Faber, London.

—— (2002) 'Flesh and Stone: The body and the city in western civilization', in G. Bridge and S. Watson (eds), *The Blackwell City Reader*, Blackwell, Oxford and Malden, MA.

Shatkin, G. (2007) *Collective Action and Urban Poverty Alleviation: Community organizations and the struggle for shelter in Manila*. Ashgate, Aldershot.

Sheller, M. and J. Urry (2006) *Mobile Technologies of the City*. Routledge, London and New York.

Short, J.R. (2001) 'Civic Engagement and Urban America'. *City*, 5, 271–80.

Sicher, E. (2003) *Rereading the City, Rereading Dickens: Representation, the novel, and urban realism.* AMS Press, New York.

Simmel, G. (1910) 'How is Society Possible?' *American Journal of Sociology* 16, 372–91.

—— (1950) 'The Metropolis and Mental Life' in K.H. Wolff (ed.) *The Sociology of Georg Simmel*, Free Press, Glencoe, 409–26.

Simpson, R. (2001) *Rogues, Rebels and Rubber Stamps: The politics of the Chicago city council from 1863 to the present.* Westview, Boulder, CO.

Sites, W. (2003) *Remaking New York: Primitive globalization and the politics of urban community*. University of Minnesota Press, Minneapolis, MI and London.

—— (2007) 'Beyond Trenches and Grassroots? Reflections on urban mobilization, fragmentation, and the anti-Wal-Mart campaign in Chicago'. *Environment and Planning A* 39, 2632–51.

Sites, W. and D. Judd (2002) 'The Limits of Urban Regime Theory: New York City under Koch, Dinkins and Giuliani', in D. Judd and P. Kantor (eds) *The Politics of Urban America: A reader*, Longman, New York, 215–32.

Skidmore-Hess, D. (2003) 'The Danish Party System and the Rise of the Right in the 2001 Parliamentary Election'. *International Social Science Review* 78, 89–111.

Skinner, Q. (1986) *Ambrogio Lorenzetti: The artist as political philosopher*. British Academy, London.

—— (2002) *Visions of Politics*. Cambridge University Press, Cambridge.

Skocpol, T. (1995) *Social Policy in the United States: Future possibilities in historical perspective*. Princeton University Press, Princeton, NJ and Chichester.

Slack, P. (1990) *The English Poor Law, 1531–1782*. Cambridge University Press, Cambridge.

Smith, A. (1776 [1843]) *An Inquiry Into the Nature and Causes of the Wealth of Nations*. Charles Knight, London.

Smith, J.M.H. (2005) *Europe after Rome: A new cultural history 500–1000*. Oxford University Press, Oxford; New York.

Soja, E.J. (2000) *Postmetropolis: Critical studies of cities and regions*. Blackwell, Oxford.

Sonn, J.W. and M. Storper (2003) *The Increasing Importance of Geographical Proximity in Technological Innovation: An analysis of U.S. patent citations, 1975–1997. What do we know about innovation?* Conference, LSE/UCLA, Sussex.

Southall, A. (2000) *The City in Time and Space*. Cambridge University Press, Cambridge.

Stein, G.J. (1998) 'Heterogeneity, Power, and Political Economy: Some current research issues in the archaeology of old world complex societies'. *Journal of Archaeological Research* 6, 1–44.

Steinberg, T. (2006) *Acts of God: The unnatural history of natural disaster in America*. Oxford University Press, New York.

Stewart, K. (1997) 'Measuring Local Democracy: The case of Vancouver'. *Canadian Journal of Urban Research* 6, 160–78.

Stone, C. (2001) 'The Atlanta Experience Re-examined: The link between agenda and regime change'. *International Journal of Urban and Regional Research* 25, 20–34.

Stone, C. and H.T. Sanders (1987) (eds) *The Politics of Urban Development*. University Press of Kansas, Lawrence.

Strom, E.A. and J.H. Mollenkopf (eds) (2006) *The Urban Politics Reader*. Routledge, Abingdon and New York.

Sudjic, D. (1995) *The 100 Mile City*. Flamingo, London.

Surborg, B., R. Vanwynsberghe and E. Wyly (2008) 'Mapping the Olympic Growth Machine'. *City* 12, 341–55.

Swanstrom, T. (1996) 'Ideas Matter: Reflections on the new regionalism'. *Cityscape* 2 5–21.

Swedberg, R. and O. Agevall (2005) *The Max Weber Dictionary: Key words and central concepts*. Stanford Social Sciences, Stanford, CA.

Swyngedouw, E. (1997) 'Neither Global nor Local: "Glocalization" and the politics of scale', in K. Cox (ed.) *Spaces of Globalization*, Guilford Press, New York, 137–66.

Swyngedouw, E. and G. Baeten (2001) 'Scaling the City: The political economy of "glocal" development – Brussels' conundrum'. *European Planning Studies* 9, 827–49.

Szelényi, I. (1996) 'Cities under Socialism – and after', in G. Andrusz, M. Harloe and I. Szelenyi (eds), *Cities after Socialism: Urban and regional change and conflict in post-socialist societies*, Blackwell, Oxford and Cambridge, MA, 286–317.

Tamari, D. (2001) 'Military Operations in Urban Environments: The case of Lebanon 1982', in M. Desch (ed.), *Soldiers in Cities: Military operations on urban terrain*, Strategic Studies Institute, Carlisle, PA.

Taylor, P. (2003) 'The State as Container: Territoriality in the Modern World-System', in N. Brenner, B. Jessop, M. Jones and G. Macleod (eds), *State/Space: A Reader*. Blackwell, Malden, MA and Oxford, 101–13.

Taylor, P.J. (2004) *World City Network: A global urban analysis*. Routledge, Abingdon.

—— (2007) 'Problematizing City/State Relations: Towards a geohistorical understanding of contemporary globalization'. *Transactions of the Institute of British Geographers* NS 32, 133–50.

Taylor, P., J. Beaverstock, G. Cook, K. Pain, H. Greenwood and N. Pandit (2003) *Financial Service Clustering and its Significance for London*. Corporation of London, London.

Therborn, G. and K.C. Ho (2009) 'Capital Cities and Their Contested Roles in the Life of Nations'. *City* 13, 53–62.

Thomas, V. (2009) 'Income Disparity and Growth'. *Global Journal of Emerging Market Economies* 1, 63–86.

Thrift, N. (2005) 'But Malice Aforethought: Cities and the natural history of hatred'. *Transactions of the Institute of British Geographers* 30, 133–50.

Thrift, N. and S. French (2002) 'The Automatic Production of Space'. *Transactions of the Institute of British Geographers* NS 27, 309.

Tilly, C. (1990) *Coercion, Capital, and European States, AD 990–1990*. Basil Blackwell, Oxford.

Townshend, C. (2006) *Easter 1916: The Irish rebellion*. Penguin, London.

Travers, T. (2004) *The Politics of London: Governing an ungovernable city*. Palgrave Macmillan, Basingstoke.

Ullman, H., J.P. Wade et al. (1996) *Shock and Awe: Achieving rapid dominance*. Center for Advanced Concepts and Technology, National Defense University, Washington DC.

United Nations (2000) *United Nations Convention Against Transnational Organized Crime*. New York.

United Nations Economic and Social Council for Asia and the Pacific (2009) 'Manila's Tondo Foreshore'. UNESCAP Macroeconomic Policy and Development Division, Bangkok.

UN HABITAT (2003) *The Challenge of Slums*. Earthscan, London.

UNHCR (2003) *Minorities at Risk Project, Assessment for Zapotecs in Mexico*. 31 December 2003, www.unhcr.org/refworld/docid/469f3ab3c.html.

Van Naerssen, T. (1988) 'Continuity and Change in the Urban Poor Movement of Manila, the Philippines' in F. Schuurman and T. Van Naerssen (eds), *Urban Social Movements in the Third World*, Routledge, London and New York, 199–219.

Varese, F. (2006) 'The Economics of the Camorra'. *Global Crime* 7, 268–73.

Vellinga, M. (1988) 'Power and Independence: The struggle for identity and integrity in urban social movements', in F. Schuurman and T. Van Naerssen (eds), *Urban Social Movements in the Third World*, Routledge, London and New York, 151–76.

Vivian, B. (1999) 'The Veil and the Visible'. *Western Journal of Communication* 63, 115–39.

Wacquant, L. (2008) *Urban Outcasts: A comparative sociology of advanced marginality*. Polity, Cambridge.

Waley, D. and T. Dean (2010) *The Italian City-Republics*. Longman, Harlow.

Wank, D.L. (1999) *Commodifying Communism: Business, trust and politics in a Chinese city*. Cambridge University Press, Cambridge.

Webber, R. (2007) 'The Metropolitan Habitus: Its manifestations, locations, and consumption profiles'. *Environment and Planning A* 39, 182–207.

Weber, M. (1966) *The City*. Free Press, New York.

—— (1968) *Economy and Society: An outline of interpretive sociology*. Bedminster Press, New York.

Weiss, M.J. (2000) *The Clustered World: How we live, what we buy, and what it all means about who we are*. Little Brown, Boston, MA.

Werbner, P. (2004) 'Veiled Interventions in Pure Space: Honour, shame and embodied struggles among Muslims in Britain and France'. *Conference on the Constructions of Minority Identities in Britain and France*, Sage, Bristol.

Whimster, S. (2007) *Understanding Weber*. Routledge, New York.

White, H.C. (2008) *Identity and Control: How social formations emerge*. Princeton University Press, Princeton, NJ.

Williams, B. (1981) *Moral Luck: Philosophical papers 1973–1980*. Cambridge University Press, Cambridge.

Williams, R. (1961) *The Long Revolution*. Chatto & Windus, London.

Winseck, D. (2008) 'The State of Media Ownership and Media Markets: Competition or concentration and why should we care?' *Sociology Compass* 2, 34–47.

Winton, A. (2004) 'Urban Violence: A guide to the literature'. *Environment and Urbanization* 16, 165–84.

Wolman, H. (1995) 'Local Government Institutions and Democratic Governance', in D. Judge, G. Stoker and H. Wolman (eds), *Theories of Urban Politics*. Sage, London, 135–59.

World Bank (2000) *Cities in Transition: World Bank urban and local government strategy*. World Bank, Washington, DC.

Worley, C. (2005) '"It's Not About Race. It's About the Community": New Labour and community cohesion'. *Critical Social Policy* 25, 483–96.

Wright, H.T. (1981) 'An Early Town on the Deh Luran Plain. Excavations at Tepe Farukhabad'. *Memoirs of the Museum of Anthropology*, University of Michigan, Ann Arbor.

Wu, F. (2004) 'Urban Poverty and Marginalization under Market Transition: The case of Chinese cities'. *International Journal of Urban and Regional Research* 28, 401–23.

—— (2006) *Globalization and the Chinese City*. Routledge, London.

Wu, F., J. Xu and A.G.O. Yeh (2007) *Urban Development in Post-reform China: State, market, space*. Routledge, London.

Yamaguchi, T. (1988) 'Recent Changes in Japan's Urban System: A Review'. *Jinbun Kagaku-ka Kiyō jinbun chiri-gaku* 10, 71–84.

Yamamoto, K. (1987) 'Regional Disparity and its Development in Postwar Japan.' *Journal of International Economic Studies* Hosei University, 132–70.

Yoffee, N. (1995) 'Political Economy in Early Mesopotamian States'. *Annual Review of Anthropology* 24, 281–311.

Young, K. (1975) *Local Politics and the Rise of Party: The London Municipal Society and the Conservative intervention in local elections 1894–1963*. Leicester University Press, Leicester.

Young, K. and P. Garside (1982) *Metropolitan London: Politics and urban change 1837–1981*. Edward Arnold, London.

Zhao, Y. (2004) 'The State, the Market, and Media Control in China', in P. Thomas and Z. Nain (eds), *Who Owns the Media?: Global trends and local resistances*. Zed Books, Penang, Malaysia, 179–212.

Žižek, S. (2008) *Violence: Six sideways reflections*. Profile, London.

Index

Page references in bold refer to sections, chapters, or parts of the book. Brackets enclosing a number after the letter n indicate a specific numbered note on the reference page – e.g. 171 (n3).